www.ingramcontent.com/pod-product-compliance
Lightning Source LLC
Chambersburg PA
CBHW022053210326
41519CB00054B/333

ISBN: 978-1-925590-30-2
Published by Vivid Publishing
P.O. Box 948, Fremantle
Western Australia 6959
www.vividpublishing.com.au

Cataloguing-in-Publication data is available from the National Library of Australia

Editing and publishing by Kelly Irving
www.kellyirving.com

Cover design and internal layout by Ellie Schroeder
www.ellieschroeder.com

www.paulbroadfoot.com

Paul Broadfoot
thinking change, accelerating **next**

X
CEL
ER
ATE

Innovate your business model,
disrupt your market and
fast-hack into the future

PAUL BROADFOOT

CON TENTS

About the author

Paul Broadfoot is passionate about what to do **next**.

As an entrepreneurial strategist, mentor, speaker, facilitator and now author, he works with enterprise executives and leaders to identify high-growth opportunities in times of rapid market change.

His fast, finite frameworks harness business model innovation to increase revenue and ignite engagement, energy and conviction for the future. Developed over years of rigorous real-world analysis and research of business disruption and innovation, these frameworks provide breakthrough market innovation.

After 20-plus years in corporate life, lamenting the lack of real conversations about business improvement, Paul set out to make a much bigger difference. Now he works with select groups of clients and companies who are serious about change, who are passionate about leading their markets and who refuse to settle for vanilla.

Paul is as comfortable developing strategy as he is talking financials. He has an MBA from the prestigious University of Chicago and a degree in chemical engineering.

He believes that the rate of change right now, catalysed by technology, is something our current generations have never seen before. That this is the reason we must learn to innovate the way we work, not just our products and services. That this is the only way 'normal' businesses will be able to thrive in the imminent future.

This book is the result of his dedication and passion to **what's possible.**

Paul will force you to think differently about your role, your organisation and your future.

More importantly, he will inspire you to act.

www.paulbroadfoot.com

Acknowledgements

Reading the acknowledgments in a book probably tells you more about the author than the rest of the book. I personally always read them. It's a great way to get the energy, passion and emotion that the author is hoping to convey to those who have helped along the ridiculous path of writing a book. (Little did I know the magnitude of the endeavour!)

So I'd not only like to thank the people who helped me become an author, but also tell them how they helped. As a result, I hope that you, the reader, will get to know me a little better ahead of the journey we are about to embark upon together.

My wife Kim – your belief and love are my major energy source. How many conversations on the couch have resulted in this book? I simply could not have done this without you.

My daughter Kimmy – you keep saying you are so proud of me: well, snap! I have always been so very proud of you.

My daughter Jacqui – you always listen to my 'business stories' with such interest and this has really helped with my thinking. You have such a curious mind, little buddy.

My mum Carolyn – for all your support, encouragement and patience.

My editor and thought distiller, Kelly Irving – without your guidance there would be no book. Your ability to clarify thinking is a talent and a gift. You helped me find my flow and get my rant on. You treated this book as your own, caring about even the smallest details.

My business manager, Melissa Angelovski – it's so much fun working with you. There is craziness, lots of laughs and loads of work! I love it when we get our high-five game on.

My mentor, Sally Anderson – I remember sitting in my backyard telling you over the phone that I wanted to do more speaking, workshops and write a book. You pointed me to Thought Leaders, and even before that, you taught me that soft skills trump hard skills. You were right on both occasions.

My teacher, Matt Church – you see the potential, the essence, in people's thinking, often more clearly than they can. The depth of your thought leadership is truly spectacular and this book wouldn't be here today without it.

My role model Peter Cook – you make everything seem so effortless, but for everything worthy, you show us how to apply our effort. You live your word, in life and in work.

My innovation advisor, Melanie Farmer – I don't think there is anyone else in Australia with as much knowledge about innovation as you. Thank you for agreeing to be interviewed and for our world-changing chats.

There are many others to thank, but a special mention goes to: Mike Herd, for your kind and wise input; Mark Hodgson and Laurel McLay, for being great TLBS mentors; Cam, for being a sounding-board extraordinaire; Ellie Schroeder, for your wonderful design work; Rosa and Sash, for the coffees and quiet place; and to my many clients, for your passion, work, belief and success.

You are all amazing.

INTRODUCTION

When did service stations become petrol stations with no pump attendant? When did they start selling Red Bull, milk, deodorant and hot cheesy wraps? When did we stop lining up at the bank to deposit a cheque? When did it become cheaper to buy new clothes than get them mended? When did TVs and appliances become disposable too? When did we stop getting landlines and buying CDs and then DVDs?

When will we know the outcome of the smartphone wars? When will we know who won the race to become lenders of choice in new-age banking?

When are we going to stop shopping at supermarkets? When are we going to stop paying to see a GP face-to-face? When are we going to stop turning lights on with our hands? When are we going to buy our first robot and a driverless car?

When will, 'When are we going to stop ...' become, 'When did we stop ...'?

If you don't start thinking about this now, then I can tell you exactly what will happen and when.

You'll lose your reputation for making the right leadership decisions, as the board loses its patience with you and your team, and your company downsizes or disappears entirely. You'll be forced to explain to your employees why their organisation – once considered to be ahead in the race – is now up shit creek without a lifejacket on.

'Artificial intelligence' will be the joke levelled not at the new robotic process automation at your competition, but at your leadership – a derogatory term for the 'old-school bosses'.

> *If you don't start acting now, then it's going to be too late.*
> _____

The world will move on. The trend will pass you, leaving you rocking on the waves in its wake, asking yourself, *'What the hell just happened?'*

Robinson Crusoe you are not. You are lumped in the same boat as all the others in big, lumbering, incumbent business today.

Maybe you're already in this position. Maybe you're already questioning why your sales are declining or asking if your future strategy is flawed.

Sure, once, 50 years ago, your business model would be good for another 50 years.

But times are a-changin', and they continue to change, *like it or not.*

The world has actually experienced this rapid speed of change before. But we weren't around in the 1700s, during the first and second Industrial Revolutions, to see that.

We weren't alive to see the first automobile roll off the ranks, to experience electricity powering our production and lighting our homes for the first time, or to travel by steam train instead of horse. Digital is doing to the world what electricity and phone communications did in the 19th and 20th centuries – only this time, the speed of change is on steroids.

Technology is changing everything, *fast.*

We need to remind ourselves that the car was once new, too.

» The advent of the telegraph enabled communication companies and killed the Pony Express.

» The advent of the locomotive enabled transport companies and killed stagecoach companies.

» The advent of electricity enabled utility companies and killed gas lantern manufacturers.

An interesting thing is happening in business right now, in the exactly same way it has happened before: new technology is enabling **new business models.**

We are seeing battles being waged not just along product innovation lines, but along **market innovation** lines. Markets are being created, expanded, shut down or shifted.

One tell-tale sign of market innovation is when companies enter spaces where they have no traditional footprint, or

when startups change the prevailing way business is done in an industry.

» The advent of the internet enabled online stores for everything and killed Blockbuster and Borders.

» The advent of driverless trucks enabled remote mining and killed jobs.

» The advent of 3D printing helped us make our planes lighter and killed some traditional manufacturing.

Don't just sit in endless meetings talking about the same old stuff, fixing the cracks that are constantly forming.

Stop fortifying your castle walls to stop startups entering. Tear down those walls and go make a new market.

Stop asking, *'Will this happen?'* and start asking, *'What must I do?'*

Unlike most business books today, this one is written from an Australian perspective (unless otherwise stated, all costs and revenue are in Australian dollars) and even provides bonus access to the first Australian research on the longevity of our ASX200 companies over the past 40 years.

It will change the way you look at your business, *now and forever*.

You'll learn:

» why your current business innovation efforts aren't leading to new customers or substantial growth

» why you need to innovate *the way you work*, not just your products and services

» how to shift the prevailing way your market operates using a fast and finite framework that identifies new innovation options easily

» how to find new growth amidst huge change.

You'll find the tools and skills you need to:

» apply a fast and finite Xcelerate framework, comprising four models, that will help you innovate your business and your market

» understand why Uber and Airbnb are Xcelerators, not disruptors, and what you need to do to be like them

» work out what type of business model you have, one of 24 current market types, in under three minutes

» get inspiration from real-life examples, stories and case studies that bring this stuff to life

» 'hack' your **market**, not just your **product.**

Many of the tools available today for business model innovation are designed for startups and focus a lot on experiential and experimental learning. This is incredibly valuable stuff and this book in no way diminishes any of that great work. This book is intended as an adrenaline injection into your thinking on the subject, so you can then use ALL the tools at your disposal.

*What this book does differently
is provide established businesses
with a proven hack to innovate
the way they work – FAST.*

—————

In many ways, this book is a precursor to the lean startup methodology that may be necessary to implement some of the outcomes. I help you achieve breakthrough thinking fast by giving you finite frameworks, not brainstorming. That's the hack.

It is not based on customer surveys, customer observation or customer testing. It is based on your company and the way your market operates. As you'll soon see, this gets us thinking about strategy immediately, and gets us very good at it. It puts your and others' strategies and business models in high resolution. This is how you create opportunity and create markets as you switch out of old ones that are about to decline.

This book is written for women and men who are here to do great work in business – for you, as leaders of large organisations, to locate and identify your current state and then look at other ways of working.

You'll learn to validate how you currently work against possible and better alternatives – ideal for working against new entrants, threats and disruptors.

My perspectives on business came from an insatiable appetite for the pursuit of better performance. I would and could never settle for average. Still to this day, I'm always looking for what's coming around the corner, and weighing up the risks. It has definitely made me a bit of a rebel at times.

In the past, I have worked as a leader in large, global B2B corporates, private companies, and with businesses of all sizes, from startups to incumbents. My thinking is refined by the distinctions and differences between the highly successful and the mediocre, from the large organisation to the small nimble business, and their respective pros and cons.

These days, I find more enjoyment consulting from the outside in. This way, I have a greater impact on many more organisations and see more transformation. I'm hired when changes need to be made, breakthroughs achieved, performances turned around or the future created.

As a corporate outsider, I am paid to get results, so I have to work out better and better ways to do that: to get the right strategy for growth, to get leaders to deepen their thinking, and to get change to stick.

I've been in those meetings where everyone is struggling to avoid bringing up the bad news, the thing that is concerning everyone, that they know is a problem. Everyone knows the unsaid gets unsorted – so I'm usually the guy to 'call it'. Yep, I'm usually that guy pointing at the big, ugly mess in the middle of the table and saying, 'What the hell are we going to do about this?'.

One of the things that happens in business, large or small, is changing when the business has to, not when it chooses to. Usually this change is forced by low-performing results. Leaders fail to react, innovate or change until it's almost too late. And if it's not too late for the company, it's too late for the leaders. If you ask my behavioural economics buddies, they'll tell you it's well studied and a very human trait. Status quo bias is in us all.

You're about to learn that longevity and performance are not like a game of pass-the-parcel at a kid's party where everyone wins and gets a prize. 'A for effort' it is not. Some companies float and others sink; some people swim and others grasp for the lifejacket, spitting water. One of our first reactions is to cut jobs. In my opinion, that's not acceptable – as leaders we've got to create jobs, not kill them.

The concepts are new, the case studies are practical (with comprehensive sources listed in the back of the book) and the real-world stories and examples help lighten the mood.

I love this quote by playwright George Bernard Shaw:

> The reasonable man adapts himself to the world; the unreasonable one persists in trying to adapt the world to himself. Therefore, all progress depends on the unreasonable man.

This is where you have to go. This book is written for you if you want to lead and enact real change.

Call yourself an innovator, a disruptor if you will, but I'm going to call you an *Xcelerator*.

The only question left to ask is – is this you?

Good.

Then let's stop asking questions and start implementing answers, and fast.

*Warning:
the freak wave
is coming*

When the automobile was first introduced, many dismissed it outright, some said it would take years and years to replace horses, and others invested in the stock and created new businesses from it.

Fast forward to today, and we're having the same conversations about robots and artificial intelligence. Many of us are dismissive, some are interested but don't see it impacting them, and others are embracing the idea and creating new businesses as we speak.

When I discuss this with clients:

» 10% are defensive

» 80% are interested but see change further away

» 10% are already out there doing something about it.

The fact is that electric lighting disrupted gas lighting, the steam train innovated past the stagecoach, the telegraph replaced the Pony Express and that was then replaced by the telephone.

There is a reason why 'disruptive innovation' has become part of everyday business language – why we are using it more and more in our meetings.

> *There is a lot for us*
> *in large organisations*
> *to be worried about.*

——————————

CEOs and execs of large corporations everywhere are frantically kicking catch-cries around boardrooms like sand on the beach:

» 'We need to disrupt our market before someone else does!'

» 'We need to innovate, *fast*, like a startup!'

» 'We need to innovate our business model!'

» 'We must act more like venture capitalists!'

» 'We must use lean startup methodology (Minimum Viable Product) for disruptive innovation!'

At least these important conversations have started in large corporations. (Tier 2 organisations have yet to catch up.)

Disruption and innovation are not enough

Disruption and **innovation** are seen as the coolest toys in the sandpit right now.

Yet these terms have been around for decades, so why are they the latest management fad now?

Why do we all, suddenly, want to play with them?

The father of 'disruption', Clayton Christensen, coined the term and the process **disruptive innovation** in 1995 in his *Harvard Business Review* paper 'Disruptive Technologies – catching the wave'. He then went on to describe the theory in detail in his book *The Innovator's Dilemma* (1997).

Most recently, in 2015, Christensen wrote a *Harvard Business Review* article entitled 'What is Disruption?', taking a thinly veiled swipe at people who hadn't read a serious book or article on the subject. He felt that things had gotten so off-track with the overuse and misuse of the term that he had to set the record straight.

Disruption is a process whereby a newer market entrant – a smaller company with fewer resources – introduces a product or service that caters to segments of the market that incumbents have overlooked or whose offering is less appealing.

The new offerings are often not as good as the incumbents, but are often cheaper.

The new offering thus finds a foothold in the market, from which the new entrant can *improve* performance to the point where an incumbent's mainstream customers start switching to the disruptor.

Branson's baby

A great example of a **disruptor** is Virgin Blue, which entered the Australian airline market as a low-cost carrier in 2000. This is exactly what Southwest Airlines did in the USA and Ryanair did in Europe.

Virgin Blue offered less service, with restrictions on meals, baggage and flexibility, in return for much cheaper airfares. This attracted a new customer market. Families that would normally drive from Melbourne to Sydney to save money could now afford to fly. Consequently, Qantas launched no-frills competitor Jetstar in 2003, which was even cheaper (with even less service).

From this position, Virgin Blue was able to expand its number of flights and introduce new services until the point they were taking mainstream business travellers away from Qantas. They became Virgin Australia in 2011.

They are a perfect example of the disruptive process.

An **innovator**, on the other hand, brings a new and superior performance offering to the market in direct opposition to an incumbent's current one and begins to take market share away.

Sometimes the new product may replace one of the innovator's existing products, or it may be introduced as a new, better and higher-priced option than their current offering, creating some cannibalisation.

Innovators typically must continue to innovate to maintain their market position. The stakes are particularly high for B2C companies with brands.

Some good examples of innovators are:

» smartphones – we continue to switch from our old models to the new ones. If it's an Apple product and we switch to a Samsung, then market share also changes

» Gillette razors – we upgrade our blades and handles, but the same faces, legs and armpits continue to get shaved

» bank loans – we switch banks based on interest-free periods or interest rates, but the overall market for loans stays the same.

Then, every so often, we see something truly impressive happen.

Like some kind of freak wave at the beach, some companies innovate and disrupt their market at the same time.

I'm going to call these companies **Xcelerators.**

And then I'm going to do something no one else has ever done.

I'm going to show you how they become Xcelerators, and how you can become one too.

Uber is not a disruptor

Table X shows some well-known examples of Xcelerators. You may know them as innovators or disruptors, but these companies are both.

Table X: Examples of Xcelerators

Xcelerator	Offering	Innovation	Disruption
		(What did they convert?)	*(Whom did they add?)*
Xerox	Photocopier	Duplicating machines and carbon copies	Spread to most offices
Apple	Smartphone	Mobile phones	Those who couldn't afford a computer but could afford a smartphone (kids, developing world)
GP2U	Virtual doctor	Traditional medical practices	Busy executives, mobility challenged, those with 'she'll be right' or 'too much hassle' attitudes
Uber	Ride	Taxis	Public transport users in a hurry
Xero	Small business accounting software	MYOB, Quickbooks	Micro businesses not using an accounting system, unsophisticated users
Google	Google Maps	Navigation devices	Melway and Gregory's users
Amazon	Amazon Fresh	Walmart, Kroger	People who used to go to supermarkets
Society One	Personal loans	Banks	People who don't like dealing with banks (is that everyone?)
Moula, Prospa, Waddle	Business loans	Banks, financial institutions	SMEs who don't like dealing with banks
CommSec	Online trading platform	Stockbrokers	Mum and dad traders, day traders
Stock Spot SelfWealth	Investing, Robo advice	Financial advisors	SMSF, share trading beginners

How to be an Xcelerator

To be an Xcelerator, you need to meet two conditions:

1. you have to add new users to the market

2. your product or service offering must be superior in performance.

Interestingly, in all the examples in Table X other than the smartphone and photocopier, the innovation was not product-based. Rather, the companies leveraged new technology to change the way the industry operated.

It's important to understand that Xcelerators are often able to both innovate and disrupt because they are **enabled by newer technology.**

The key word here is **enabled**. They did not **invent** the technology.

Companies that **invent** technologies are likely to become innovators in their fields (like Bose Corporation, which you'll read about in 4: Wired for Sound).

Companies that **use** the new technologies have a great shot at being Xcelerators, like those listed in Table X.

GPS services integrated into smartphones; **live bank feeds** for Xero; **online platforms** for share trading; **algorithms and the internet** for investing companies.

In periods of technological change, we get more **market innovation** – that is, a change in the way an industry operates, not just a product or service innovation.

Grab a surfboard
or get wiped out

If you work in a large organisation then you must strive to become an Xcelerator; otherwise, you will not move fast enough or be able to surf with your new competition.

You'll just get sucked up and washed out to sea.

You are the incumbent, and incumbents struggle to disrupt themselves.

But you have the scale, resources and muscle that startups typically don't have. So you don't want to just disrupt your market, you want to Xcelerate it.

Large organisations have greater inertia than smaller ones. You have more to lose and more history to anchor you, but encouragingly, you also have more resources to re-deploy.

If you aim to be an Xcelerator, you must ensure that you morph and move as the freak wave erodes the sand from under your feet.

It is not enough to be a disruptor. It is not enough to stay an innovator.

Teams of well-funded startups
are not just riding the waves;
they are creating them.
They are coming for you.

———————

The only thing you can do to survive is to innovate **and** disrupt – you need to **Xcelerate**.

According to Accenture, global FinTech investments are growing at 75% and exceeded US$22 billion in 2015. Fuelled by technology, much of this FinTech activity will bring better products to market, be preferred by customers and be cheaper.

Different, better and cheaper is how you Xcelerate.

The tools provided in this book are designed to help you to become **Xcelerators**. You will learn how to ***innovate your business model*** and ***disrupt your market*** at the same time.

So stop playing with the sand – the surf is up!

*Let's learn how to
create the wave
and Xcelerate
your business.*

———————

Part X

UNDER
ATTACK!

Every day we are hit by new rapid changes that will affect our daily lives – from China's first 3D-printed home to Amazon Go's checkout-free shopping and Elon Musk's SpaceX trip to Mars. We are seeing and experiencing what once was the stuff of sci-fi movies. Fiction is turning into reality!

Our wildest dreams are coming true. We are getting more of what we want, when we want it, through a stunning stream of new technology. But there is a flipside to all of this, and it's more than a little sinister.

The changes that make consumers happy are starting to make others lose their jobs. Those organisations that are NOT participating in this revolution are under increasing strain. It is a fact that the lifespans of our largest, most successful global organisations are getting shorter.

As you're about to learn in Part X, fortifying your defences and turning your back on the imminent waves of change is a dangerous, perilous stance.

Our largest businesses, with decades of optimisation and efficiency-led thinking, now seem distinctly old and slow. As someone in a position of forward-thinking leadership, you cannot deny this. That's why you are spending hours in meetings discussing innovation and disruption.

Only one thing is certain – new times call for new ways.

Australian organisations in particular have scant track record of successful innovation. We are languishing behind global standards. We have little evidence we are ready for the pace of change we need to stay relevant. We are in desperate need of a fast hack.

You need to get your head in the game, the tools in your hands and shift your thinking on how to create a future business.

Part X will show you why.

1

**Walls
don't
work**

1

He had one blue eye and one brown. He washed his hair in saffron to keep it orange and had a phobia of cats. He was king at 20 and dead by 32. During that time, he founded 70 cities, ruled the ancient Macedonian region, never lost a battle as a military commander and somehow managed to squeeze in teachings from Plato and Aristotle.

We are talking about a leader with a legacy, known for strategy, redrawing his market and innovation.

We are talking about Alexander the Great.

So what can we learn about business today from a dead king?

In their time, Alexander and his army made a mark on this world like some could say Steve Jobs and Apple, Jeff Bezos and Amazon and Bill Gates and Microsoft have in modern times.

One of Alexander's most widely known claims to fame is the Siege of Tyre in 332 BC. He had just defeated the Persians at the Battle of Issus and King Darius III of Persia had fled towards Asia.

Alexander was faced with two choices:

1. pursue Darius deep into Asia

2. ensure Phoenicia and Egypt were under his control and didn't represent a threat to Greece before pursuing him.

Alexander chose the latter. His intention was to conquer the city of Tyre, the key naval port of Phoenicia.

At the time, Tyre was part of the powerful Phoenician civilisation along the Mediterranean Sea. Although the old city was on the mainland, the new city of Tyre was built on a three-kilometre island of rock further fortified with towering walls.

The Tyrians were great traders and had a formidable navy, which concerned Alexander. Despite the burgeoning reputation of the young king and his military prowess, Alexander's initial messenger was refused entry to the city. Had he been invited in, the city would likely have been occupied without further bloodshed.

The Tyrians were clearly confident that Tyre's walls would hold up.

Alexander was always relentless, never idle and never one to give up. He was confident of his skill with strategy and tactics, which had won him many a battle so far.

So his strategy to attack Tyre was to build a mole, or causeway, out to the island. This was no small undertaking: more a feat of engineering. The causeway had to be 800 metres in length from the mainland out to the island and wide enough to accommodate his 40,000-strong army.

How did the battle play out?

The army made good progress with the causeway, building it as they went, until they hit deep water nearer the island. They were constantly under fire from Tyrian ships, and the closer they got to Tyre the more they were fired upon by catapults and archers stationed on the walls of the city.

Alexander built two siege engines (towers) at the end of the causeway to return fire and add some protection for his men. But the work became even more difficult as Phoenician divers sabotaged the work underwater at night.

In the first major turning point in the siege, the Tyrians loaded an old horse transport boat to the gunwales with flammable naphtha. They towed this toward the siege towers and set it alight, as they swam back to shore.

The siege towers were destroyed. Round 1 to the Tyrians.

Alexander, never one to be outdone, immediately ordered the rebuilding of the causeway to double its original size. Recognising he was at considerable disadvantage without a navy, he took a small contingent back to a previously captured city, Sidon, and commandeered their boats, about 80 of them. The king of Cyprus also threw his lot in with Alexander, sending another 120 boats.

Alexander now had a navy three times the size of the Tyrians'. Round 2 to Alexander.

As the siege continued, many innovative strategies were employed by both sides.

The Tyrians

- » sent divers to cut the anchor ropes of Alexander's ships

- » heated sand and dumped it over the city walls, using the wind to carry it and burn their enemies' clothing and ship sails

- » hid their southern port via a vast array of sails to conceal a naval counterattack

- » moored boats in a line across the northern harbour entrance

- » reinforced some ships with armour

- » hurled enormous rocks into the shore waters to prevent boats getting closer.

Alexander

- » flaunted his entire 200-ship navy across the northern horizon to deter a drawn-out naval engagement

- » tested the Tyrian defences with small attacks by boat

- » mounted siege engines on ships and began pounding the walls

- » used his ships to ram Tyrian ships and the city's walls.

How did this all end? And what does this teach us about modern business?

The modern business battle

The Tyrian wall to the south was repeatedly rammed by Alexander's ships, while the Cyprians created a diversionary attack from the north. The decision to use ships to ram the walls of an island castle was quite an innovation. Alexander led the breach of the wall and plundered the city of Tyre.

Final round – Alexander.

In our history to date, there has never been a wall built for protection that has continued to successfully protect its community, city or country. All of these walls – Hadrian's Wall, the Wall of Constantinople, the Great Wall of China, even the Berlin Wall – have at some time been breached either physically or diplomatically (for example, a castle's surrender).

And yet we still continue to build them. Why?

Castles were once thought to be the state-of-the-art strategy in medieval warfare. But even one of the world's most famous and well-preserved castles today, Rochester Castle in Kent, England, fell in the First Barons' War of 1215. Castles as a military strategy were superseded from around that time.

The Siege of Tyre shows us how difficult it is to pursue a defensive strategy and win. Most builders and defenders of castle cities believed they were safe.

In our modern world, our business is like a walled castle where we often feel safe from assault. But this is proving to be less and less so. As you'll soon discover, companies aren't lasting as long as they used to. Redundancies have become an organisation's way of making the numbers when the strategy isn't strong enough.

Why aren't we making a much bigger deal about redundancies as a sign of poor strategy? It's as though we have become accustomed to thousands of people losing their jobs.

Alexander was **on the move;** his was the insurgent army, not the incumbent walled city. Everyone he fought fell. He had no castle.

Have you ever been in meetings where your competition or a new trend in the market was dismissed as no threat, with nothing to be learned?

It's not an uncommon sentiment, especially when past performance has been good for our corporations, to be at best complacent and at worst cocky.

Hunkering down behind your company's walls, rather than getting out there and leading changes in how your market works, will inevitably lead to your waving the white flag.

———————

Weapon versus strategy

Historically, as we saw with the siege of Tyre, there are two ways armies would innovate to defeat enemies:

1. bring a new **weapon** to the battle (e.g. the navy)

2. employ a new battle **strategy** (e.g. make a causeway, ram the walls, create a diversion, test for weaknesses).

Our modern-day corporate equivalents are:

1. bring a new **weapon** to the market (e.g. innovate a new product/service)

2. employ a new battle **strategy** (e.g. innovate the *way* we work).

Leaders of consumer product companies such as Samsung, Apple, Sony, P&G, Nestle, and 3M must continually find ways to innovate their weapons because they are in the business of building consumer products and brands. They are B2C companies and must continue to innovate and bring these new weapons to market.

B2B businesses typically do less of this. Their markets are generally a bit slower to move – **but herein lies the danger for both types of businesses.**

If you're focused on innovating your weapon and someone changes the way of working, the way business is done in an industry – if they innovate their business model and their market instead of their product – their strategy will kill you.

The product innovation problem

To understand the difference between the two, let's look at a classic Apple example:

1. when Apple launched the iPod, they launched a new product – aka a new **weapon**

2. when Apple enabled music downloads, they changed the prevailing way the music industry worked, thus creating an entirely new battle **strategy.**

It is becoming increasingly common to see new market entrants disrupting our large corporations. With rapid technological change, we will see more and more market innovation through the way companies operate in their markets.

We must stop thinking only of innovating our products and services and start innovating the way we work.

———————

Many of today's most successful companies have not invented anything; they are just changing the way things are done.

1. **Webjet** changed the way we book flights – they didn't invent passengers, planes or the internet

2. **Xero** changed the way small businesses keep their books with automatic data feeds and monthly subscriptions – they didn't invent small businesses, accounting or the internet

3. **Netflix** changed how Australians pay for TV and movies, forcing Foxtel to adjust – they didn't invent movies or downloading from the internet

All three examples above were *technology enabled,* but their primary innovations were *the way business is done.*

Under siege!

Some of the world's largest and most successful companies are under siege right now.

Consider Microsoft, Apple and IBM. In Australia, it's Australia Post, Foxtel, and our Big Four banks. Soon it will be our supermarkets.

Microsoft has already lost several battles:

» Internet Explorer is gone (which at one point occupied 95% of the market) – Microsoft limps on with Bing and Microsoft Edge, but the internet search battle has been lost to Google

» mobile operating systems (Windows Phone and

Windows mobile) – this battle was lost to Apple (iOS) and Google (Android)

» cloud services – this battle was lost to Amazon Web Services (AWS).

At time of print, Apple's share of the smartphone market is declining for both handsets and operating system. IDC reports Android has 88% of the operating system market, and in the handset market new entrants Xiaomi, OPPO and Vivo are rapidly growing from China, whilst South Korea's Samsung maintains market leadership.

Some of these companies have well and truly lost the battle to survive.

» Blockbuster filed for bankruptcy in 2010 after losing out to Netflix

» Angus and Robertson (2011), Collins Booksellers (2005) and Borders (2011) lost out to Amazon

» Sharp was acquired by Foxconn in 2016, thanks to unsustainable debt and market share lost to LG and Samsung.

So ask yourself whether you feel safe; whether you're happy with your results right now. Are you?

Hmm, that's what the Tyrians thought too.

Today's battle strategy

When under siege, a company has limited ability to focus on innovating new weapons or strategies whilst also repelling attackers.

Relocating the castle is often not an option, and at some point, when survival is actually threatened, we may be forced to leave our old cities and start again. Just like IBM did when they left the PC market after pioneering it.

Yet this can be an arduous task that most, if not all, organisations struggle to achieve.

It is rare to see a successful and established business creating its own future before it is challenged – and doing it well.

But this is what is needed in our latest industrial revolution: the digital revolution. New technologies are enablers of new ways of working, not just of new products, and thus we see tremendous disruption.

Not enough time is spent innovating the way a company works: its business model, its revenue model, the way it differentiates, and the way it goes to market.

Innovating prevailing norms in markets is the new battle strategy.

———————————

Some traditional businesses have managed to adapt, albeit painfully.

Xerox, GE and IBM are enduring organisations that have innovated their business models in order to survive over 100 years. They often did so during periods of great pressure caused by eroding markets and terrible financial performance. All have had near-death experiences in their histories.

Startups and smaller disruptive organisations have a distinct advantage over established companies. They are not guided or anchored by their past; they are focused on their futures and they have fewer assets to be redirected if a new business model is needed. And we all know where they sit on the hungry versus complacent spectrum.

We could cite and recount countless battles in history that show how superior forces were overcome by a smaller opponent using a more advanced military strategy.

Today it's happening as startups become 'unicorns' (private startups with at least a billion dollar valuation).

Amazon went online selling books in 1994 (from Jeff Bezos's garage: a cliché pioneered by so many successful startups). Collins Booksellers passed into administration around ten years later in 2005, as did Angus and Robertson and Borders in 2011. With over ten years for the Australian booksellers to react, counter, adapt, innovate and change, one might think it would have been possible to fight their way out of declining profitability.

Often long-term trends are very evident. The incumbents can see them coming. Kodak had 30 years' notice of the changes wrought by their own invention, the digital camera, before succumbing in 2012. Telstra had some very ugly times with

landline decline before turning it around. Australia Post is in the throes of reinventing itself now, but declining numbers of posted letters didn't just creep up and say 'Boo!' Our Big Four banks and others are grappling with the rise of FInTech firms.

All these trends are inexorable and permanent. They don't represent a cyclical downturn like the mining industry. They are permanent shifts. They are not turning around.

We knew landlines, snail mail, book and DVD sales in stores would all shrink rapidly.

So why then weren't the organisations successful in moving far enough and fast enough?

There is clearly more to this than just being aware of the shifts occurring. And to argue that management teams have not worked to adapt is too harsh. It's difficult terrain.

But clearly there is a need to do much more, and faster.

How is your castle doing?

A Boston Consulting Group project, which researched 30,000 public firms in the US over 50 years, found that:

» public companies have a one-in-three chance of being delisted in the next five years

» the average age at which companies disappear has almost halved over the past 40 years

» the rate at which companies are disappearing is six times higher than 40 years ago.

Another study, Innosight's *Corporate Longevity: Turbulence Ahead for Large Organizations*, predicted that 50% of current

S&P 500 companies would be replaced over the next ten years. In 1965, a company was expected to be in the S&P 500 for 33 years. This is now forecast to be less than half by 2026.

More than half of the Fortune 500 companies listed in the index for the year 2000 have disappeared already. That means the lifespans of the largest corporations in the US are shrinking.

And what about in Australia?

How are organisations in Australia doing today? And what does this mean for you if you are a leader of one?

You're about to discover this in 2: Australian research leaves a bitter taste. I reveal the results of a study I commissioned with Dr Andrew Pratley of the University of Sydney. This is the first time these findings have been made public. But first, remember that ...

It's a contact sport, not a chess game

With corporations disappearing faster than ever, and with the digital revolution upon us, we need to:

» ensure we have a portfolio of business at different growth stages – mature, growth and startup

» focus on innovating the market – the way the industry operates (not just new products), as do many of the startups

» embrace and leverage changes in technology.

This is an active pursuit, not a passive one. It's a contact sport, not a chess game. It's a science experiment, not a mathematic theorem. It's about diversity and collaboration for better creation.

All these require fast action.

Diversity matters

Perhaps one reason why we have less adaptation and considered shifts in our strategies at a corporate level is the lack of diversity in many organisations.

One element of this is the lack of gender diversity in our senior executive ranks. In the 2015 *Diversity Matters* report by McKinsey, they 'examined proprietary data sets for 366 public companies across a range of industries in Canada, Latin America, the United Kingdom, and the United States'. They studied the correlations for board and management diversity with financial performance. Here are three of their findings:

» Companies in the top quartile for racial and ethnic diversity are 35% more likely to have financial returns above their respective national industry medians.

» Companies in the top quartile for gender diversity are 15% more likely to have financial returns above their respective national industry medians.

» Companies in the bottom quartile both for gender and for ethnicity and race are statistically less likely to achieve above-average financial returns than the average companies in the data set (that is, bottom-quartile companies are actually lagging, rather than merely not leading).

By its very nature, diversity creates different perspectives, viewpoints, opinions, analysis capabilities, emotions and alternatives. This is why you can't stay in the same business model for too long.

As he moved through Asia, Alexander the Great came to realise he needed to adopt and integrate some of the cultures of the lands he had conquered.

Partha Bose, author of *Alexander the Great's Art of Strategy,* goes into further detail about how simply winning the battle does not win the battle. Hearts and minds, being open and sensitive to customs and social norms, and learning were all things Alexander picked up.

In many ways war is a terrible metaphor, in terms of what it represents, but it has strong parallels to a corporation striving for leadership of a market. There are winners and losers. The best thing about winning in business today is that it's starting to matter *how* we get there. But just like diversity, we're taking a hell of a long time to get it right.

The sharpest tool

Japanese electronics giant Sharp was over 100 years old when it was taken over by Taiwan's Foxconn in 2016, following massive debt.

Its claim to fame was the invention of the mechanical pencil in 1915. The Ever-Sharp gave the company its name and spawned several innovations and achievements such as the first commercial transistor calculator in 1964, the LCD calculator in 1973 and the first camera phone in 2000.

From 2005 to 2010, Sharp was the market share leader for mobile phones in Japan. For a time it was also one of the leaders for flat screen TVs globally. It was the first company to release an 80-inch LCD LED TV (in 2011) and was the first to make a robot phone, called RoBoHon, that used artificial intelligence and could do everything a smartphone did, plus act as a projector, recognise faces when taking pictures and greet people it recognised.

Despite such a proud history of product innovation, even Sharp didn't make it. Erosion of market share was one of their downfalls. What a shame for everyone involved.

It's a timely reminder that you are never too big to fail, no matter how sharp you think you are.

The corporation castle conundrum

Enterprises tend to be large, fixed-asset cities, which makes it difficult to travel too far from the centre of mass. Safety lies within the walls. The thinking often stays within the walls too. (Google 'group think' as a cognitive bias.)

Companies often have a proud history, an emotional attachment to a place or community and a city to defend. They have often invested decades in establishing and building their fortress. This is where a corporation's traditional size advantage is proving a weakness.

So what to do?

It's time to focus on the future and not hunker down behind our walls.

If we aren't dismantling our own castle walls and setting out on expeditions to new lands, then we might as well await the inevitable siege headed our way.

Xcelerate now

1. Traditional corporations are like modern-day castles – they're asset-heavy and slow to change.

2. Hunkering behind your organisation's walls is like waiting to be besieged and then waving a white flag to insurgent disruptors. You may feel safe and secure, but you are not.

3. There are two ways to win a business battle: a new weapon or a new strategy. Large companies focus too much on innovating their next product or service (their weapon) and not enough on the way they work and their market (their strategy).

4. To be an Xcelerator, you need to innovate how you work.

5. Market innovation is a battle strategy too rarely seen in the modern day.

2

Australian research leaves a bitter taste

2

My grandfather worked for Tooth & Co, a brewery in Kent, Sydney, for 50 years. The stock market crash of 1929 forced him to leave school in Form three (Year nine) and get a job. He was 15 when he went to work there and he retired when he was 65. Apart from a four-year stint in WWII, where he got a grenade fragment in his chest, he remained working at the brewery.

The fact he stayed with the same company for 50 years says something about my grandfather, but it says more about the times. It's a hard number to fathom in the 21st century. Nomadic careers are more commonplace than ever before.

Tooth & Co isn't around anymore. The organisation was delisted from the ASX 200 in 2010 after 175 years. This is one of the ways a company exits the ASX 200; it begins to struggle and ends up purchased by another organisation.

Founded in 1835 by John Tooth and his brother-in-law John Newnham, Tooth & Co was one of Australia's oldest companies. Two of their beers have somehow managed to survive: KB Lager and Kent Old Brown.

At one stage the company was so successful it even owned Penfolds Wines and the NSW Hungry Jack's franchise.

Before the 1970s, beer sales were a territorial affair with states having handshake agreements not to distribute into each other's territories. Hotel licensees were contracted to buy exclusively from the breweries that owned them. Both practices were outlawed in 1974, which was the beginning of the end for many entities as competition increased, putting sales and margins under pressure. Consolidation began.

Some other well-known names that have disappeared from the ASX are:

> Wattyl, Greyhound Pioneer, Ampol, Harris Scarfe, Travelodge, Yellow Cabs, Pioneer, John Sands, Vulcan, Australian Newsprint Mills, Katies, Sebel, Tooheys, Foster's, Acmil, WC Penfold, Darrell Lea, Braemar

In many cases, these brands live on with new parents.

These companies have been delisted from the ASX 200 due to either buyout or failure. That's how the corporate lifecycle goes, right?

Well, let's see.

Bitter endings

A controlling interest in Tooth & Co was acquired by David Jones in 1981. David Jones at the time was owned by corporate raider the Adelaide Steamship Company.

The Adelaide Steamship Company operated with huge debt levels and when the crunch of the early 1990s recession came, it began large asset divestitures.

In 1983 Tooth & Co was sold to Carlton and United Brewery (CUB), with previous combatants VB (Victoria Bitter) and KB Lager now in the same stable.

The Adelaide Steamship Company, under its new name Residual Assco Group, was delisted in 1999.

Newton's First Law

Most companies behave like the objects that Sir Isaac Newton used to study in his laboratories. That is, they tend to move in a straight line unless acted upon by an external force. That's Newton's First Law of Motion.

From an outside view, it always seems that time and again companies either:

1. fail to take sufficient and substantial action to adapt to changing times until they face a watershed moment when their very survival is threatened or when their CEOs or boards have failed to provide even mildly acceptable performance; or

2. suffer substantial changes to their industry that are outside their control (e.g. a major regulation change).

In the former, more common occurrence, there is a tendency for organisations to stay their course, even when their markets may be changing. So what happens during industrial revolutions when landscapes change more rapidly?

The logical conclusion is that those companies most affected by any market change will be impacted more quickly. Presumably, some will either fail or, more likely, they'll be acquired, either voluntarily or from a hostile takeover.

It follows that the rate of delisting would be higher in times of faster change.

But that's all just guesswork now, isn't it? So to test our hypothesis, we crunched the numbers.

What hasn't been analysed in Australia until now is whether these patterns are changing. No study of this kind exists in

Australia – so we set out to change that.

What you're about to read are hard statistics on the delisting rates of companies. We use the delisting rate as a proxy for company longevity.

The following research involved studying 5270 companies that had been delisted from the ASX 200 between 1975 and 2015. We mapped the delisting rates for organisations since the advent of reliable data in 1975.

The research we commissioned was conducted by Dr Andrew Pratley and Harlan Ikin of the University of Sydney Business School. We analysed the *maximum length** of time (in years) that a company was in business before being delisted. (*N.B.: We corrected for name changes. For example, if a company changed its name we took the time for delisting as the highest possible number based on this. If a company was delisted and relisted, we combined the totals. We termed this the 'maximum' delisting age.)

The results are shown in Figure 2.1.

Figure 2.1: Company delisting age by decade

Decade	Average maximum delisting age
1975–1984	36.9
1985–1994	17.2
1995–2004	15.4
2005–2015	15.2
Average	18.6

Looking at the data in decade segments, we see the average maximum delisting age has declined by more than half over that period: from just over 35 years to just over 15 years.

Did you get that?

I'll say that again.

Company lifespans in Australia have halved over the past 40 years

That is an epic transformation!

This means more change, more failure and more mergers and acquisitions.

The other interesting piece of information from Figure 2.1 is that the decline in delisting rate seems to have slowed in the most recent decade. Has most of the disruptive change arrived or is something else going on? Is there enough innovation to create new competition?

The ASX 200 represents our biggest companies. Figure 2.2 shows some more summary statistics from the study.

Figure 2.2: ASX 200 delisting summary statistics

Key Figure	Result
Average delisting age	18.6 years
Median delisting age	11 years
1-year delisting rate	14.8%
5-year delisting rate	44.4%
10-year delisting rate	71.0%
90% delist within	23 years

There are some very surprising numbers here. The first is that 44.4% of our largest companies are disappearing after five years. That is not dissimilar to the SME space, which is often quoted as having a 50% fail in five years.

The other surprise is that the average (18.6 years) is far higher than the median (11 years). This is due to our very largest corporations such as Westpac (founded in 1817) bumping up the averages. Our largest companies are very old. How are we doing on new large companies entering our upper echelon?

This is embarrassing

Each year the Australian Government's Office of the Chief Economist publishes the 130-plus page *Australian Innovation System Report.*

Australia consistently punches well *below* its weight regarding the OECD innovation measures.

The December 2016 report states, 'After matching to OECD definitions of business size and sector, the data suggests that Australia is not an innovation leader but an innovation follower.'

Worse still, our larger organisations rank 29th out of 30 OECD countries. The report states:

> Compared to 31 other OECD countries, Australia, at nine per cent, ranked 23rd for the year 2012–13. This ranking largely reflects the activity of SMEs (with 10–249 employees) at nine per cent. Large businesses (with 250+ employees), also at nine per cent, rank even lower at 29th out of 30 OECD countries.

If you take a look at our top 20 ASX 200 companies by market cap, you'd note two things. Firstly, they are very old – they average 99 years, with Westpac the oldest at 200 years. And the second thing you'll note is the lack of any new companies that look anything like a 'unicorn', a disrupter, or an innovator.

There are no Facebooks, Googles, Apples, Amazons or Microsofts in Australia's Top 20

Competing forces

How do we reconcile our two key points?

1. Our top 200 companies are delisting younger that at any other point in the past 40 years, but that rate of decline has slowed.

2. Our large businesses rank 29th out of 30 OECD countries on innovation.

One possible conclusion is that the digital revolution hasn't fully arrived in Australia yet. That previous declines in company lifespans are a result of cheap funding, M&A activity, globalisation and increased competition, and big breakthroughs from us leveraging technology for new business models is yet to be tapped. Is our woeful innovation performance shielding our incumbents for now?

Ask yourself:

» Can you see our companies innovating their markets?

» Are we launching global businesses?

» Can you see examples of companies launching new types of businesses by utilising new technologies?

» Will our future competition come from overseas because we are too slow? (Imagine Amazon Go against Coles and Woolworths, Netflix against Foxtel, Global FinTech against our banks.)

*Disruption, innovation,
change, speed, technology,
globalisation – call them
what you will, but they
are all conspiring to shorten
company lifecycles,
and the innovation game
is passing us by.*

*C'mon, Australia
– we are much
better than this.*

———————

3

*King tides
of change*

3

The humble sewing machine entered estates and households in the 1850s, reducing the time spent mending garments from several days to a few hours.

Singer Manufacturing Corporation's Scottish factory added 7000 workers in its first five years. Sales of sewing machines grew exponentially over two decades to two million units by 1876.

The industrial application of the sewing machine resulted in large factories employing thousands of workers to make clothes. Making a shirt was reduced from 10–14 hours to only a single hour, spawning a new wave of mass production.

This resulted in a couple of things. It increased supply and reduced costs. The reduced costs and increased supply in turn led to another two things: demand increased and selling and distribution commenced. Export markets were created. This led to even more demand, which turned people's attention to finding more workers and obtaining better productivity. Several things were needed to make all this work.

We needed:

- » a **system** of work (enter the corporation!)

- » **labour** (pulling people from the country to the city, and from the fields to the factories)

- » **capital** (to build the factories and pay the workers – again, cue the corporation with equity raised from shareholders and investors).

And so companies grew in size, funded by investors and shareholders. They were focused on productivity to keep up with the rate of growth of demand. This capitalist era emerged from more agrarian, feudal ways of life.

The rise of capitalism sought ever-increasing ways to mass-produce and innovate productivity. As time went on, the focus on productivity went from (1) keeping up with demand to (2) being a weapon against competition to sell at lower prices and then, as company markets began to decline, (3) productivity was needed to stay economically viable.

Then and now

The technological changes of the first and second Industrial Revolutions were breathtaking. First we saw water power, then steam, then mechanisation, mass production, electricity – and straight after that, globalisation. Everything was working to reduce manufacturing costs and make things more efficient and cheaper.

What would it have been like if you'd lived back in those days? What would you have thought the first time you heard about electricity, a telephone, a steam train, a car?

It would sound incredible, right?

Maybe it would have been like how we felt when we first heard about artificially intelligent robots, biometric sensors on our bodies that report our health, driverless cars or 3D-printed houses ...

Yet this is now our reality!

With each technological advancement came greater time-saving, improved standards of living (on average) and often more factory workers for the new products.

During their initial rise, back then, corporations needed more of the same, but better and cheaper and faster.

Large corporations enjoyed the competitive advantage of being able to scale their labour and asset pools more easily with better access to both.

You're about to read what is changing now. These changes are starting to undo the traditional advantages that the corporation had. It's the 'big guns' that have some distinct disadvantages, both now and into the future.

Revolutions wrap-up

First Industrial Revolution
(late 1700s to early 1800s)

Brought water and then steam power, which was the start of less manual labour in some jobs, as well as pulp mills, textiles and metal-beating. The second half of the 1800s rounded out the first Industrial Revolution with steam-powered transport and early mechanisation.

Inventions: Steam engine, spinning jenny and flying shuttle for textiles, telegraph, steamship, threshing machine, bicycle, metal lathe, steam locomotive, Fourdrinier paper machine, battery, typewriter, sewing machine, reaper, electric dynamo, mechanical calculator, first revolver, rubber, steel, and the electric printing telegraph (fax) machine.

Second Industrial Revolution (late 1800s to early 1900s)

Ushered in electric generators, mass production and the production line, as well as railroads, steel production, rubber, telegraph, light bulbs, the internal combustion engine, steamships and papermaking.

Inventions: *Dishwasher, rayon, combustion engine, rotary washing machine, plastic, dynamite, barbed wire, four-stroke engine, phonograph, telephone, moving pictures, photography, cash register, steam turbine, machine gun, automobile, AC motor, transformer, escalator, motion pictures, vacuum cleaner, radio, air conditioner and aircraft.*

Back to the digitised future

The current digital revolution is undoing a great deal of the outcomes of the first and second Industrial Revolutions. With access to capital and labour more democratised and accessible, there is a fundamental shift away from the industrial era's framework.

The music industry, for one, has morphed a number of times and is representative of what changes when technology changes. Digital music downloads in the US are declining at double-digit rates. Streaming is killing digital downloads, which is rather ironic given that quite recently the digital download killed the CD (which in turn killed the LP and cassette).

Digital marketing of digital music is producing stunning results for the artists embracing it. Nielsen's mid-2016 music report shows streaming up 58.7% and digital downloads from the likes of iTunes down 18.4% for albums and 23.9% for tracks.

Business models for the music biz have changed yet again. This time, the money is made not from music sales but from subscription-based advertising. Once music turned digital, it was always going to morph.

Of course, a lot more than just music has gone digital. So what does this mean if you're a large organisation today?

Any corporation that wants to have a future must get better at fast market shifts.

Change, collaborate and create need to replace systemise, optimise, rinse and repeat. It's not just about better, cheaper, faster; it's about **different**, better, cheaper, faster.

There are a number of king tides hitting corporate beaches right now. These tides have the potential to erode the foundations of large fixed-asset organisations. Conversely, for businesses that aren't anchored to their structures (like startups) these king tides can have the opposite effect and help them float to the top.

This is cause for concern if you're a large organisation today. Many of the tides do not play to the advantages of scale, size, position or history. You need to become aware of the tides – not so you can seek higher ground, but so you can run down the beach to jump in and surf them.

So what are some of the king tides you need to be aware of?

King tide 1:
access is less privileged

A research study into the use of mobile phones and the internet in Cambodia found the following:

- » 94% of Cambodians own a mobile phone

- » 39.5% of these owned at least one smartphone

- » 15.2% of those with no formal education owned a smartphone.

And this major study was a year ago now.

With a smartphone we have the internet, and with the internet we have Google, enabling us to access knowledge. YouTube,

Khan Academy and Google have a tremendous array of learning covered.

It won't be long before the first Cambodian with no formal education launches a startup that alters the way a market works. And it's likely they'll launch it from their smartphone.

Or perhaps they'll just start by helping out someone else's open innovation effort, like M. Arie Kurniawan from Salatiga, Indonesia, who designed a jet engine bracket in a global 3D printing challenge held by GE. Kurniawan had no aeronautical engineering experience and was just 19 years old. He upstaged 700 other contestants, solving a problem that had defied GE's engineers.

The idea that access is less privileged doesn't end here. Previously, to get funding for your startup, specialist skills for your business, advice, know-how or access global markets, you had to know people, have connections or expend significant effort and money finding and attracting what you needed.

Now we have:

» Kickstarter – crowdsourced funding and community building

» AngelList – online startup funding

» Kaggle – data scientists analysing big data

» eBay – global niche sales beating local market sales

» Khan Academy – free education

» Local incubators and accelerators pooling shared expertise and providing access to mentors

» Upwork and Fiverr – services helping people build collateral to promote their businesses from as little as five bucks.

All these services provide access to business-building tools.

We are further optimising how we 'operate the planet' and our participation in it.

Global empowerment, financing and labour markets are reducing the relevance of national boundaries, increasing global trade and building capabilities where there were previously none. We are even starting to see pushback in the political arena in the US and Europe against the early outcomes of some of these trends, which mean less work for some.

So we will see more change, from different places, and vastly increased innovation. New Silicon Valleys will pop up on a smaller scale, everywhere.

Changes in markets will come much more swiftly, and for those companies facing adverse changes, only the nimble will beat the decline in average lifecycle.

King tide 2:
abundance of everything

The collaboration and sharing economy, as it has been dubbed, is facilitating access to goods and services that have *not* traditionally been offered at scale. When this happens, **additional capacity is brought online.** In economics, when additional capacity comes online prices decrease and competition increases.

Here are just a few examples of our sharing economy:

- » homes offered to strangers as accommodation – Airbnb

- » homes offered to strangers' dogs as accommodation – DogVacay

- » home car spaces offered to strangers – Parkhound

- » cars offered to strangers for rides – Lyft, Jayride, MyCarPooling, La Mule, Flexicar, Shebah, Uber

- » cars offered to strangers when not in use by owners – Getaround, RelayRides (now Turo)

- » bikes offered to strangers when not in use by owners – Liquid

- » expensive assets like camera equipment and kitchen processors offered – Simplist

- » shed gear offered to strangers to use or try – Open Shed

- » cooked meals shared with strangers – HomeDine and LeftoverSwap

- » leisure gear shared with strangers – Spinlister

- » wifi networks shared with strangers – Fon

What does this mean?

1. There is an additional supply of capacity in many areas facilitated by technology. During the initial stage, prices will drop, companies that can't compete will suffer sales decline and demand will increase, driving even more suppliers to enter the market. When demand and supply reach a new equilibrium, some incumbents will be gone and many new companies will be in the market.

2. Traditional businesses with large investments in asset use for sales are being disrupted. They face price and demand pressures. And just like Blockbuster, the more bricks-and-mortar they own and the higher the fixed costs, the more tenuous their chances of adapting. Yes, new demand (more transactions due to lower prices) will enter the market also, but then so will additional supply (additional competition). Essentially, this latent capacity being brought online will be great for standards of living everywhere, but will also be tougher on traditional asset-heavy businesses. Air travel is very likely to be disrupted by new business models, as it is an extremely asset-intensive industry.

3. The trends in business will be more toward rental models than ownership models.

4. There will be an entrepreneurial spike, which will likely end in overcapacity and consolidation of both new and traditional businesses.

5. To win you will need the best service, the best prices or the best products – being average just won't cut it.

King tide 3:
turning into a 2-year-old

We all know how demanding two-year-olds can be – they have no patience, they are determined to get their own way and they push boundaries to see what is possible (for them). They are egocentric: purely focused on themselves.

Now think about how we watch TV.

Once upon a time, we grew up watching our favourite programs when they were scheduled. If we had to work late, travel or dine out, we missed the show.

But now we can binge on what we want, when we want it.

I want what I want, now, **on demand, for me**, just like a two-year-old.

Once we become used to what we want, delivered very quickly and tailored for us, then traditional businesses must adapt the ways they service their customers.

The more your service offering can provide speed and tailoring, the more modern your business will be.

Your website needs to deliver dynamic content to me.

Your product must be available to me when I am ready to buy.

Your service needs to adjust to my specific preferences.

Period.

King tide 4: don't just stick to your knitting

Some businesses are building new capabilities from their platform of technologies. Where we used to define companies by the industries they played in, we are increasingly seeing changes in technology facilitating companies into new market segments.

This is particularly true when products and services go digital. They become eminently more transportable when they are digital. It's often easier to 'manufacture' and sell a digital product than a physical one.

The following businesses are no longer defined by the markets of their initial triumphs:

- » Apple has made a US$1 billion investment in Chinese ride-hailing app Didi Chuxing

- » Google has entered the transport and delivery business with driverless cars and drones

- » Amazon builds products: Kindle and Echo.

So what does this mean?

- » It is far harder to predict who your competitors will be. When shifts occur in your markets, in many cases the underlying source of change may not be obvious. All you may notice is declining sales in one of your market segments, which you may put down initially to the competition's pricing or the economy or sales and marketing performance. Sub-currents, ripples and erosion will be prevalent but not necessarily obvious.

- » The correlation between new products and higher price points will be broken: new products will be better AND cheaper.

- » Increasing pressure on product innovation (B2C) companies to match the global pace of every potential competitor – big and small.

Take your medicine

McKesson is one of the standout B2B companies in a sea of B2C companies at the top of the Fortune 500. It has over 70,000 employees globally and revenue of over US$180 billion, coming in at number five in revenue on the Fortune 500 list for 2016 behind Walmart (1), Exxon Mobil (2), Apple (3), and Berkshire Hathaway (4).

Incredibly, McKesson is 183 years old. It started in 1833 as Olcott and McKesson, named after its two founders, who began wholesaling botanical and herbal products in New York City.

The company is still in the business of wholesaling. They still sell prescription and over-the-counter drugs and other products, but they have also moved their business model into retailing, distribution and a myriad of health services.

Not only have they witnessed incredible changes to the regulatory health system in the US, they have been one of the innovators by adopting and rolling out technologies along the way.

As early as the 1980s, one of their businesses, McKesson Drug Company, was receiving 99% of their orders electronically from their massive network of

retail pharmacy customers. Customers would place these orders and receive their orders the next day at a 99% service level for prescription drugs and 93% for over-the-counter items.

Some of their other services included:

- » management reports containing everything from advertising incentives from drug companies to inventory levels and sales data by product

- » credit cards systems, terminals

- » consulting on location and store design for new pharmacies

- » a management system for hospital buying groups.

As the world moves again into a period of rapid technology advance and the health landscape changes once more, McKesson is focused on moving with the times, as it always has.

In a 2015 interview with Fortune Magazine, CEO John Hammergren said:

'We're always focused on disrupting ourselves inside of McKesson and in inventing the future for our customers.'

McKesson has a technology division that helps its customers leverage technology in their businesses, providing services like hospital management systems.

Already using big data and analytics, the healthcare industry is changing quickly. Patient records are being automated at a rapid rate and there is a joint focus with other industry players on making IT systems and communications between doctors, hospitals, pharmacies, patients and even machines much more interoperable. At the moment there is still a lot of work to do in this area.

'Personalised medicine' is evolving. Doctors can access data and analytics to review the most successful surgical methods, best treatment protocols based on the specific disease and on the individual. This goes a long way toward avoiding over- and under-medication too.

McKesson has started a venture capital business to partner with entrepreneurs and VCs in order to understand where health and technology are headed and to assist their customers with those transitions. Healthcare is complex and highly regulated. It has also been a highly competitive industry over the decades. Getting future trends right means big bets on technology.

The IT systems McKesson introduced in the 1980s saved their customers time and money, enabled their sales force to value-add rather than pick up orders, optimised internal costs and productivity, reduced inventory for the pharmacies, improved service levels, helped with pricing and specials and pooled all parties' buying power. Over 12 years (1975–1987), McKesson's sales in this department grew by 424% and their costs by 86%. And their customers were better off.

They'll be aiming to bring a bit more of that type of medicine to customers in this digital revolution.

So if you want to lead a long-serving company like McKesson, you'll need to stop waiting for the king tides to wash over you and start implementing the tools in this book. Part Y will show you how to do that.

Xcelerate now

1. Huge demand was created when industrial revolution goods could be made faster and cheaper. Corporations were formed to provide access to capital, labour and systems of work that could fill this demand.

2. The digital revolution is granting access to capital and labour without needing a large organisation – this is undermining the traditional strength of an enterprise.

3. Today's business leaders and most businesses were not around during the first or second Industrial Revolutions, which means they have difficulty seeing how fast things are changing and how quickly we need to respond.

4. There are major trends affecting your future market. You need to be aware of these king tides:

    a. Access for all, to all – education, capital, labour.

    b. An abundance of everything – prices are falling, competition is increasing.

    c. Buyers expect services to be on-demand and tailored for them.

    d. Businesses are jumping traditional industry walls and competing based on technology platforms, rather than staying in their historical markets.

5. Companies that embrace new technology to enable new ways of working will survive and thrive. Will you be one of them?

Part Y

———

STRATEGY

So now you know why you have to disrupt your own thinking and learn to innovate the way you work if you want to be a successful organisation that stays in the game.

Part Y is designed to show you how to do that.

The Xcelerate framework that we're about to walk through comprises four models:

1. business
2. revenue
3. communication
4. differentiation.

Each of these four models is a ***finite framework*** – a known set of options – that is designed to allow you to quickly see where you're at now, and easily determine where you could be.

You will discover the 24 different types of business models available in today's market, and you'll learn to work out which one you currently have and how to generate new options for your organisation. You'll also learn how to generate potential options in each of the other three models, too.

Unlike a lot of innovation that starts with ideation or problem solving and generates divergent options deliberately to get to a creative space, these tools operate the opposite way initially. They converge your thinking. They provide a strategy fast-hack.

They are ***very*** powerful. You will immediately start to operate at a strategic level.

Fast is often associated with less rigour – but not so here.

In this case we have taken care of that rigour for you. We've done the heavy lifting to get to this point: countless hours of consulting and countless hours of research into the successful and unsuccessful.

After each of the models is a case study of a successful and well-known organisation that will help bring each model to life and show you how easy it is to apply the Xcelerate framework. They are your inspiration!

Part Y will muster the belief and conviction you'll need to leave the castle, breach the walls and break away in a new direction.

So let's get to work designing your future strategy.

4

*Business
model*

4

My very first Uber trip was an early morning pick-up to the airport, quite a while ago now. I had no idea what to expect before I hit the 'Request uberX' button. *What car was coming? How long would it take to arrive? How much would it cost? Would it even turn up?*

When my driver arrived (at the exact time, like the app had told me), I received a bottle of water on the way and even had a pleasant chat with the driver. Over the countless times I have used this service, I've met lots of colourful characters: a mum who had just dropped her kids at school, a businessman who used to be a major liquor importer into Malaysia, a retired advertising exec who was too bored at home and a surfer making some extra bucks. It was a far cry from the usual wait-and-see-if-my-cab-is-gonna-turn-up-today type of ride I'd been accustomed to.

Uber is heralded as the poster child for 'disruption', and more specifically as a technology disruptor. It connects drivers to riders through an app in 540 cities worldwide; it connects us with a service and a bunch of interesting people all at the

touch of a button on our smartphone.

That's product innovation, right? A new weapon right?

Wrong.

Uber is a great example of a company that rocked a complacent industry with a superior *and* cheaper experience by changing the prevailing way the industry worked by introducing a new business model.

Remember, not all startups invent some kind of revolutionary technology.

They often just use technology to facilitate a new business model.

———————

As we discussed earlier, Uber is more than a disruptor – it's an Xcelerator.

The trouble is that the term 'business model' is now ubiquitous and we've lost any firm mental resolution as to exactly what its definition is.

Is it our strategy? Is it how we create and capture value for customers? Is it how we are different to our competition?

So I want you to forget everything you've been told about what a business model is.

We're going to flip the traditional business model definition that you might read in the dictionary and bring a new definition into play.

Down to business

We know that Netflix, Amazon, Xero and Uber all operated very differently to Blockbuster, Angus and Robertson, MYOB and Yellow Cabs when they entered their markets. They are new technology startups, right?

Wrong, again.

When Amazon started selling books over the internet, Jeff Bezos was still at his kitchen table. He didn't have the skills, money or need to invent a technology to take his new business model to market. He just leveraged the internet.

Netflix didn't invent the DVD when they started as a mail-order business in 1997. Xero didn't invent cloud computing in 2006 when they launched. Uber didn't invent location-based services in smartphones.

All these businesses had exponential growth curves and were Xcelerators. All of them changed the way their industries worked. All of them had a dramatically negative impact on incumbent businesses.

We know they are 'different', but we need to diagnose exactly what they did so that you – as an incumbent – have a chance to do something different too.

An MIT Sloan paper drafted in 2006 explored the performance of the USA's 1000 largest businesses in a paper entitled 'Do Some Business Models Perform Better Than Others?'. Part of this study posed the question, 'What is a business model?' To work out the answer, the authors assessed what 'rights' were sold and what 'assets' were involved in the different organisations.

This approach to classifying a business model enabled very clear distinctions. Since then, I have built upon and changed the asset types to better reflect current times and newer business models.

My simple formula for working out a business model

Ask yourself the following two questions:

1. What is the income-generating **ASSET**?

2. What is the income-generating *activity*?

Together, the **ASSET** type and the *activity* type make up an individual business model.

So what are the types of **ASSETS** and *activities* available in the market today?

☆ The really important stuff to bookmark

The types of **ASSETS** that cover today's market are shown in Figure 4.1.

Figure 4.1: Types of income-generating ASSETS

ASSET	Description	Example
PHYSICAL	Products, buildings	Groceries, cars
FINANCIAL	Shares, cash, loans, insurance	Home loans, life insurance
DIGITAL	Digital files	Digital music, digital movies
KNOWLEDGE	Theories, applications, patents, education, trades, accreditation	Universities, plumbers, doctors
MARKETPLACE	Places where people come to buy and sell. Easy to list products for sale, easy to make purchases. Price is determined by the market (buyer and seller) rather than an intermediary.	Local markets, eBay
SYNDICATE	Member groups, social networks, associations	Industry associations, Facebook

The types of **activities** in today's markets are shown in Figure 4.2.

Figure 4.2: Types of income-generating activity

activity	Description
distribution	An **ASSET** passes into ownership and is then sold to another party (without major alteration).
connection	Two or more parties (often buyers and sellers) are introduced for them to transact. Ownership of the **ASSET** is not taken. Options are common.
creation	Value-adding is generated via combining inputs into a transformed finished product.
contracting	Temporary use of an owned/controlled **ASSET** is allowed/offered, after which it can often be re-tasked.

This is the really important stuff you need to know, and you'll keep referring back to it long after you've read this book.

We will continue to describe business models as **ASSET:** *activity*.

All the examples that follow rely on this info so bookmark it, tag it or earmark this page now.

The 24 types of business models

Given the above, there are 24 types of business models available in the market today.

The best way to view this is to put them in a table – Figure 4.3.

Figure 4.3: The 24 types of business models

ASSET TYPE	activity			
	distribution	connection	creation	contracting
PHYSICAL				
FINANCIAL				
DIGITAL				
KNOWLEDGE				
MARKETPLACE				
SYNDICATE				

Now let's add in some company examples to see what their business models are, shown in Figure 4.4.

Figure 4.4: Business model examples

ASSET TYPE	activity			
	distribution	connection	creation	contracting
PHYSICAL	a supermarket		a manufacturer	a landlord
FINANCIAL				a bank
DIGITAL	Foxtel movies			
KNOWLEDGE				a doctor
MARKETPLACE		eBay		
SYNDICATE		LinkedIn		

It's sometimes not easy to classify enterprises with lots of business units as having just one business model. In this case, working at business unit level is more helpful.

Putting your own business in the table makes it easier to see where the 'blanks' are. Some of those will represent breakthrough business ideas, either for you or for your competition (if you are too slow).

Airbnb under the microscope

Let's work out an example together so we can put it into the table.

Example 1
Airbnb – online accommodation

QUESTIONS

1. **What is Airbnb's income-generating ASSET?**

   Airbnb is a **MARKETPLACE** where buyers check out sellers' rooms. It is possible for a buyer to evaluate different options based on many criteria and then make a choice. The seller is free to present different aspects of their property, location, rooms, interaction and features.

2. **What is Airbnb's income-generating activity?**

   Airbnb is facilitating a **connection** between a buyer (guest) and seller (premises).

Uber – not what it seems

Often, until we stop to work out the different business models of organisations they can seem very similar.

Uber, like Airbnb, is held up as an example of a modern tech startup in the sharing economy that connects people.

One party has spare capacity (in their car or home) and the other party would like to use that. But there are important distinctions beneath the surface.

Companies that seem very similar are often very different.

Want to see?

Example 2
Uber – ride-sharing

QUESTIONS

1. What is Uber's income-generating ASSET?

Uber provides drivers, vehicles and rides. However, it does not own the vehicles and it does not employ the drivers, so the 'rides' are the **ASSET**. (Taxis, on the other hand, need a licence to operate so this is their asset.) The

---

rides are a **PHYSICAL ASSET** because they rely on a car and driver to provide the ride.

2. What is Uber's income-generating activity?

In other words, how does Uber make money from its asset, the ride?

» **Creation**? – No, since they aren't transforming the car/driver combination in any way.

» **Connection**? – No. Although they are connecting two parties, the element of reviewing different options and choice is missing.

» **Distribution** – No. They are not 'on-selling' the asset, as it gets used again.

» **Contracting**? – Yes! Uber is granting use of the asset and after use, they are free to contract it out again.

ANSWER: Uber's business model is **PHYSICAL**: *contracting*

So let's add Airbnb and Uber into the table so we can compare them, Figure 4.5.

Figure 4.5: Business models – Airbnb and Uber

ASSET TYPE	activity			
	distribution	connection	creation	contracting
PHYSICAL	a supermarket		a manufacturer	a landlord Uber
FINANCIAL				a bank
DIGITAL	Foxtel movies			
KNOWLEDGE				a doctor
MARKETPLACE		eBay Airbnb		
SYNDICATE		LinkedIn		

These two companies are often seen or described as being the same, but they have very different business models.

Startup swap

What would it look like if you swapped Uber and Airbnb's business models?

Scenario A
It's late as you get out of your last business meeting for the day. You've already checked out of your hotel, since the deal was supposed to be done and dusted by lunchtime, but now you need to stay an extra night.

It's too late to ring your office and ask them to set up another night's accommodation, so you walk out of your client's office, grab your smartphone and hit 'Request AirbnbX now'.

Up pops a message, the closest room is booked and it gives you a map to walk to your hotel. You already had all your preferences and criteria stored in the app.

This accommodation solution now has the same business model as Uber – PHYSICAL: *contracting* (just like a hotel).

Scenario B

You're on the train home, after a big night out in the city.

It's late, and you don't feel safe walking on your own from the train station, after drinking. So you want to book a trustworthy and experienced female driver to give you a lift.

You open up your ride-sharing app and type in a few preferences. You're given a number of alternatives, complete with reviews and ratings.

You select Lucy, who promises to put on relaxing music and has a few comments saying she locks the doors when she drives at night for safety. By the time you get off your train, she will be waiting for you.

This transport solution now has the same business model as Airbnb – MARKETPLACE: *connection* (just like Stayz).

Swapping a business model
can be done easily and quickly
– in under three minutes.
We'll walk through this
in 12: Diving in.

———————

Putting more into practice

What other business models could we figure out?

Let's have a look at three businesses involving connection of two other parties: Aussie Home Loans, SEEK and eHarmony.

They are all 'brokers', so they have the same income-generating activity – ***connection.***

But what is each company's income-generating **ASSET***?*

Aussie Home Loans sells mortgage loans, so it has a **FINANCIAL** asset.

SEEK's income-generating asset is the online **MARKETPLACE** where job posters and job seekers go to find each other.

eHarmony sells relationships, right? Could we define that as a **MARKETPLACE**? Well, yes (although it's not always easy to determine who the buyer and who the seller is in the relationship), but their **ASSET** is their claim to 'match' people. eHarmony's founder Neil Clark Warren, a psychologist and author, worked his whole career before putting together a model for compatibility for relationships. His matching service for singles was all about finding people who were compatible with each other, not only people who shared the same interests. So eHarmony actually provides a **KNOWLEDGE ASSET**.

Therefore, these examples have the following business models:

Aussie Home Loans – **FINANCIAL:** *connection*

SEEK – **MARKETPLACE:** *connection*

eHarmony – **KNOWLEDGE:** *connection*

Let's put them into the business model table to see how each of these compare, Figure 4.6.

Figure 4.6: Business model comparison

ASSET TYPE \ activity	distribution	connection	creation	contracting
PHYSICAL				
FINANCIAL		Aussie Home Loans		
DIGITAL				
KNOWLEDGE		eHarmony		
MARKETPLACE		SEEK		
SYNDICATE				

Download a more detailed
table of current businesses
and their business models at
paulbroadfoot.com/resources

Australia's ostrich problem

A typical Australian manufacturing firm has the following business model: **PHYSICAL: *creation.***

Imagine you are the CEO of a manufacturing firm. You've just completed your annual planning and concluded that your market is undergoing a permanent shift to imported products. A tipping point has been reached.

If that threatens your main revenue source, what can you do?

You have five options:

1. reduce costs in line with reducing sales or reducing profitability

2. find a new market

3. find a new product

4. increase sales effort or skill to gain new customers (increase your share of a shrinking pie)

5. bury your head in the sand and hope things improve (aka: the ostrich strategy).

You have a manufacturing plant with hundreds of employees whose jobs are on the line.

Yet I see option five happen a lot in business.

It's easy for executives faced with poor performance to rationalise away the results as an impact from the economic cycle, the competition reducing prices, upcoming elections, a handover from the financial crisis or poor staff. The delusion of thinking 'It will turn around' is so alluring.

The first four options are OK, but probably won't change much if the trends against you are permanent (in any industry).

Figure 4.7 outlines some options for new business models.

Figure 4.7: New business model options for manufacturing

ASSET TYPE	activity			
	distribution	connection	creation	contracting
PHYSICAL	A		Manufacturing	
FINANCIAL				
DIGITAL				
KNOWLEDGE			C	B
MARKETPLACE		D		
SYNDICATE				

Option A: PHYSICAL: *distribution*

What if you found some lower-cost manufacturing options overseas, set up a distribution agreement, preferably exclusive, and become an importer? Businesses tend to underestimate the value of their services, believing that it's the product alone that gives them their business.

Option B: KNOWLEDGE: *contracting*

If your knowledge specialties are in the application of products in customer environments, you may have a strong offering. Or what if you could handle the import, customs clearance, supplier negotiations, monitoring and management of the supply chain? Developing a service business around either of these could work.

Option C: KNOWLEDGE: *creation*

In this case, you would package up your expertise in your game and collect, collate and codify it to sell to others in your industry: franchising, licensing or servicing, for example.

Option D: MARKETPLACE: *connection*

Perhaps it isn't easy for buyers to find suppliers in your industry. Or perhaps it is difficult to compare products, because everyone has a different specification. So why not build a **MARKETPLACE** where buyers can review multiple products and suppliers in a common format? Conversely, it may not be easy for suppliers to contact their target market, and this may be a way for them to do so.

Top 6 challenges

Large businesses often have massive inertia and lots at stake. It's no wonder these six key challenges make us hunker down behind our castle walls:

1. **short-term view** – *quarterly reporting to shareholders*

2. **angst over employees** – *jobs loss and skill adjustments to new businesses*

3. **sunk costs** – *people have trouble getting their heads away from asset and employee investments already made*

4. **lack of conviction** – *that a new model can work*

5. **risk aversion** – *cannibalisation, failure, human nature*

6. **understanding information** – *consistent frustrations or wants of customers*

With the pace of change increasing, however, we will see more large enterprises, as well as much less visible mid-tier organisations, forced to shift their business models to survive.

That's why now is the time to embrace the change!

High performers learn to lead markets

The world's best organisations and performers have made business model changes **before** they have been forced to.

Apple, Netflix and Amazon are good examples of this.

Let's fill them into a table so you can see their different business models, Figure 4.8 (overleaf).

The fact that Apple, Netflix and Amazon have great cash flow and are willing to sustain losses to establish markets has been one reason for their ability to stay out in front.

But it would be a cop-out for us to use this as an excuse to avoid leading in our spaces.

These leaders have chosen to take additional risks – because the status quo is riskier.

Let's take a look at a few of the risks these companies faced.

- » When Netflix added movie downloads, they risked cannibalising their existing mail-order customers.

- » When Amazon invested into physical assets (warehouses), they were changing their totally on-line presence to a more traditional one.

- » When Amazon launched Kindle eBooks, it posed a serious cannibalisation threat to their existing sales.

- » When Apple launched Apple Music and Apple iTunes, they were moving away from their history as a B2C product innovator.

- » When Apple launched Apple Music, they knew it would cannibalise iTunes sales.

- » When Netflix launched their original content they had NO experience in this business, which requires huge up-front costs, picking winners for audiences, creativity, production and so on.

As Steve Jobs himself once put it:

'If you don't cannibalize yourself, someone else will.'

———————————

Figure 4.8: High performers and their business models

	activity			
	distribution	connection	creation	contracting
PHYSICAL	Netflix (mailorder DVDs) Amazon (books) Amazon Fresh Amazon Go		Apple Mac Apple iPod Apple iPhone Apple iWatch Amazon Data centres Amazon Kindle, Echo	Amazon (fulfillment)
FINANCIAL				
DIGITAL	Apple iTunes Netflix (SVOD) Amazon (eBooks)	Apple Music	Netflix (original content) Amazon Web Services	
KNOWLEDGE				
MARKETPLACE				
SYNDICATE				

Musical MARKETPLACES

How do you determine if you have a **MARKETPLACE ASSET** *or not? Let's look at the music industry to work this out.*

The three big music-streaming options are Spotify, Apple Music and Amazon Music Unlimited. We can also still buy and download songs on iTunes or buy CDs on Amazon and eBay, not to mention watching YouTube clips.

For a business model to have a **MARKETPLACE** *as an* **ASSET,** *it must be:*

1. *somewhere people go to trade: that is, both buy and sell*

2. *somewhere people can buy and list products for sale relatively easily*

3. *somewhere the market (the buyer and seller) determines the price, as opposed to an intermediary.*

The music streamers, not the music artists, set prices for the buyers, so this falls short of criteria (3) for a **MARKETPLACE***. Spotify, for example, pays artists using an algorithm based on market share of 'plays'. Music streamers tend not to take ownership of the music (although this is complicated). Suffice to say, Spotify is either* **DIGITAL: *distribution*** *or* **DIGITAL: *connection,*** *depending on whether they take ownership or not.*

CD sales are **PHYSICAL: *distribution*** *and Apple iTunes is* **DIGITAL: *distribution.***

The power of
Facebook and LinkedIn

Facebook and LinkedIn are based on membership and community. They are **SYNDICATES** – that is, a group of individuals or organisations combined to promote a common interest.

Their power lies in the size and engagement of their *membership.* Were this to fall, so would their performance as organisations.

As the world becomes more what I call 'Triple C' – collaborative, creative and crowd-sourced – online **SYNDICATES** will proliferate.

A **SYNDICATE'S** customers are its members. The more it focuses on serving its members and helping them build community, the more significant and stable a **SYNDICATE** becomes.

A spotlight on SYNDICATES

Here are some further examples of **SYNDICATES**:

SYNDICATE: *connection*
Facebook, LinkedIn The income-generating *activity* of these businesses is about connecting people. That is the reason their members joined.

SYNDICATE: *distribution*
GrabCAD, Workbench GrabCAD is an online community of engineers whose purpose is to share all the tools needed for Computer Aided

Design (CAD). Workbench is a free cloud-based collaboration solution where people and companies can work together to design new products and engineer solutions.

These communities' primary goal is the ***distribution*** of information, education, tools, and insights to their members, keeping them informed so they can do great work. Both these organisations would argue they also foster ***connection,*** but it's a much weaker element than the primary one.

SYNDICATE: *creation*

Kickstarter, Melbourne Angels

These types of **SYNDICATES** are growing rapidly. They thrive on a sense of purpose. Kickstarter is a crowdsourcing community whose mission is to help bring creative projects to life.

Kickstarter is a great example of a successful **SYNDICATE** that helps artists, musicians, filmmakers, designers, and other creatives find the resources and support they need to make their ideas a reality. To date, tens of thousands of creative projects – big and small – have come to life with the support of the Kickstarter community.

Melbourne Angels and the other Angel investing communities similarly help their members invest in, typically, startup businesses that have plans to grow.

XChange (Business in Heels), TradeYa

Business in Heels, a 50,000-strong network of businesswomen across the world, has introduced XChange, where their members can go to trade (or purchase) each other's services.

'Business in Heels executive director Lisa Sweeney said she and her co-owner Jo Plummer discovered that running networking events was not a practical solution to generating business opportunities between businesswomen.'

This is an example of a member network group working to help its members contract business services to each other.

Similarly, TradeYa is a modern take on an old-fashioned barter system where people trade services to each other.

Possibility power

Business models are now no longer the mystery they used to be. These days you have 24 clear options – 24 choices of business models.

It is quite straightforward to identify which business model you have now and what other models you could move to.

There is real power in exploring these possibilities, as you will see from the next case study.

Xcelerate now

1. A business model answers the two questions:

   i.   What is the income-generating **ASSET**?

   ii. What is the income generating *activity*?

2. There are six **ASSET** types and four *activity* types, which constitute 24 different business models available in the current market.

3. This fast, finite framework for business models gives us a fast-hack to breakthrough ideas.

4. Some businesses often held up as shining examples of tech startups have:

   a. actually not invented anything

   b. different business models to each other, in many cases.

5. Innovating your business model reveals how to shift your markets and challenge prevailing norms.

5

**Be first
and be
lonely**

5

Ginni Rometty has been with her organisation for 35 years. She was 24 when she started there.

During an interview with *Fortune* senior writer Jessi Hempel at Fortune's Most Powerful Women Summit, Ginni said:

» **'Growth and comfort do not coexist'**
(personal growth occurs when you are challenged and take some risk despite your inner critic)

» **'Diversity is about inclusion'**

» **'Be first and be lonely'** to make a new market

» **'What do you believe?'** is a great question, instead of 'What's your focus?'

Ginni also **doesn't believe in the inevitable (you can change things).**

And, by the way, she's the CEO of:

> » the oldest technology company in the world – 105 years old

> » the company that pioneered the PC for consumers and invented LASIK eye surgery, magnetic stripes on credit cards, ATMs, floppy discs, and the scanning tunnelling microscope capable of viewing and manipulating individual atoms

> » an organisation with over US$80 billion in annual sales.

Ginni Rometty runs IBM.

IBM has adjusted its **business model** in a BIG way a couple of times. In the early 1990s IBM found itself in many markets with outmoded and rapidly commoditising products. It reached its lowest point when IBM posted the largest quarterly corporate loss in history at the time: US$5.5 billion.

IBM (or International Business Machines) started as a tech product inventor and manufacturer. **PHYSICAL: *creation*** was their business model, and they were very good at it. Three things are very demanding for this type of company – the constant need for invention and commercialisation, feeding the monster (being big means the wins need to be very big too) and the fact that being in tech makes those first two issues bigger, because the market shifts quickest without the benefit of strong product brands (like food or pharma, for example).

In 1991, IBM had a new strategy (and new business model) ratified by the board. They would launch a business called IBM Consulting Group, which later became IBM Global Services. This would move IBM from a hardware and software company to

one that derived a substantial part of its income from services – the **KNOWLEDGE: *contracting*** business model.

Lou Gerstner was brought in as CEO in 1993, and as he documents in his book *Who Says Elephants Can't Dance?* (2002), there was no better than a one-in-five chance that they would survive. Many at the time thought Gerstner would continue the trajectory of breaking the company up into independent operating units. Significantly, mainframe sales, IBM's leading product line and source of most of its profits had collapsed from US$13B to US$7B in a scant three years. Revenues had declined from US$69B (1990) to US$62.7B (1993) and the combined losses over the three years 1991 to 93 were US$15.9B. The onset of this drastic decline was as surprising as it was swift.

From 1980 to 1990 the company grew from US$26B to US$69B, having introduced the wildly successful personal computer in 1981. Lou Gerstner's turnaround of IBM was nothing short of stunning. Revenues returned to growth and the company to profits. Revenue grew from US$62.7B in 1993 to US$87.5B in 1999. But what happened in the intervening period around their business model switch? In 1992, when the new business model was launched, services revenue was under 10% of total revenue. Today it is around 60%.

Back to Rometty. The April 2016 quarter showed IBM's lowest revenue for 14 years. At time of writing, IBM had posted 18 consecutive quarters (4.5 years) of declining annual revenue. Rometty finds herself presiding over a company not dissimilar to the one Lou Gerstner found before him in 1993. Technology has evolved so quickly that again IBM, slow to adopt the cloud, has been hit. It is again their original business hardware-based model (**PHYSICAL: *creation***) that

is creating the pressure. Once again, we hear cries that IBM is doomed: a lumbering giant of yesteryear.

So what is Rometty doing about it? Rometty detailed her plans to reboot IBM in the *Wall Street Journal* (2015):

- » championing cognitive computing with IBM's artificially intelligent Watson computer (the one that won *Jeopardy* and bested the world's chess champion)

- » making large investments in cloud centres and its cloud platform

- » focussing on data analytics and big data to create a new market

- » bringing mobility to enterprise.

So Rometty is going all-in on some of the digital revolution's key platforms: cloud, artificial intelligence and big data. She is wagering the company, her reputation and her tenure on leadership into new technologies and creating new markets with them.

I like it. I like it a lot.

She is shifting, in part, to a new business model of **DIGITAL: *creation***, whereby the Watson AI, data analytics and security, mobility and cloud solutions are deployed into businesses around the world.

IBM has always been at its best when leading, or as Rometty puts it: ***'Be first and be lonely'***. This will be a test for her and this venerable company that has by sheer weight of history beaten all the odds. Will the launch of this latest business model be another turnaround in the making? It's too simplistic by far to suggest that business model changes

are the only impact on turnarounds. What we can say is:

» Changing a company's business model is difficult, and is often only done in times of challenge and poor performance.

» If IBM had retained only their initial **PHYSICAL:** *creation* business model to the present day, it is very unlikely they would still be in business.

» Your business model is only one aspect of how you operate – and it's not always the thing you need to change, as we will see later.

» Changing your business model requires great nerve, risk and discomfort.

So remember Rometty's words:

*'Growth and comfort
do not coexist'.*

_____

Revenue model

6

Once upon a time, mum or dad might have yelled at you for staying on the phone too long. You were told to keep it short when speaking to Grandma interstate, or you had to wait until 7.30pm to call your friends because it was cheaper.

Back then, you might have paid a monthly bill for your landline with a total for each 'transaction' made. Local calls, interstate calls and mobile charges were specific costs, with some attracting a higher charge based on distance or duration or both. There were peak and off-peak rates. You paid for what you used, according to the fee schedule. No more and no less.

These days, you have a phone plan with allowable limits on calls. Many plans have unlimited calls and include SMS and MMS, and you pay the same amount every month. Plan choice now emphasises how much data usage you are allowed before you're charged more (at ridiculous rates). These data limits often determine which network provider you choose. You don't pay for the actual phones up front; it's all part of the monthly charge.

Our lives are a list of subscriptions. Just think about how you pay for your phone, internet, gym, movies, and software.

Who is responsible for this? Well, we are.

Many of us convert to subscriptions to receive newer products and services, especially digital ones. There can be lots of advantages, such as not having to remember to pay the bill each month and spreading the cost out over the year.

Subscriptions represent a trend.

And swimming with or against a trend presents your business with an **opportunity**.

Turn your model into money

You can turn your business model into money by changing your revenue model.

Changing your revenue model is a lot easier than changing your business model.

It is also a great opportunity to increase a business's growth rate with existing clients, to increase profitability, or appeal to new customers in a different segtf the market.

Companies have often increased their fortunes significantly just by changing their revenue model:

>> **Xerox** moved from selling to leasing photocopiers in the 1960s

>> **Microsoft** moved from selling licenses in their software to subscriptions

>> **Google** moved to a pay-per-click auction model.

The result of Xerox and Google changing their revenue model was exponential growth. That's not a figure of speech. I am talking about sales hitting an exponential upwards growth curve (like an Xcelerator).

To work out a revenue model, ask yourself the following two questions:

1. Will the customer **own** or **rent** your **ASSET**?

2. Are there **three** parties involved, or just **two**?

Your four revenue models

There are 4 different types of revenue model you may have, depending on how you answer those two questions (as shown in Figure 6.1):

1. TRANSACTION

2. INTRODUCTION

3. AFFILIATION

4. UTILISATION

Let's walk through each in more detail.

Figure 6.1: Your four types of revenue model

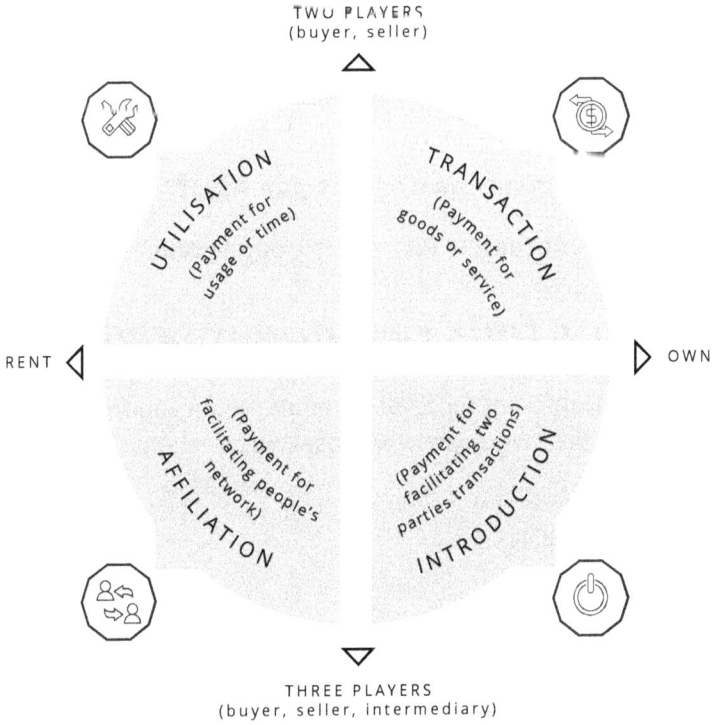

TWO PLAYERS
(buyer, seller)

UTILISATION
(Payment for usage or time)

TRANSACTION
(Payment for goods or service)

RENT

OWN

(Payment for facilitating people's network)
AFFILIATION

(Payment for facilitating two parties transactions)
INTRODUCTION

THREE PLAYERS
(buyer, seller, intermediary)

1. TRANSACTION *revenue model*

This is the most common revenue model. A TRANSACTION involves only **two parties.**

A TRANSACTION is the sale of a product or service. If it is a product TRANSACTION, ownership changes from one person or organisation to the next. If it is a service TRANSACTION, then the service is often paid for once completed.

There are many examples of this type of revenue model: supermarkets, trades (such as plumbers), consumer goods, taxis, flights, industrial products and house purchases.

> TRANSACTION
> Two parties involved, ownership changes,
> clear beginning and end

2. INTRODUCTION *revenue model*

The INTRODUCTION revenue model still involves a change in ownership and the sale has a clear beginning and end, but there are **three parties** involved.

Real estate agents, brokers and recruiters often operate this model.

So too does eBay. They earn a fee for listing and then a percentage once the item sells.

Incidentally, Pierre Omidyar, founder of eBay (or AuctionWeb, as it was originally called), didn't start charging for use of his platform until his internet provider contacted him, saying he needed to move to a business account due to the traffic his

site was receiving. That move raised his monthly internet fee from US$30/month to $250/month.

> **INTRODUCTION**
> Three parties involved, ownership changes,
> clear beginning and end

3. *UTILISATION* **revenue model**

A UTILISATION revenue model involves payment in exchange for usage, time, access of some other unit. Ownership does not typically change and there are just two parties involved.

In this way, we are granted access to software like Microsoft Office and antivirus programs, gyms, online news, pay TV and office spaces.

Xero 'rents' their software to users for a **subscription fee** each month. Ownership of the asset never changes, and they sell directly to the subscriber so there are only two parties involved.

This forced MYOB to respond to Xero's move into their market. They changed their revenue model from selling licenses to a similar monthly subscription. The user pays for the UTILISATION of the software each month.

Typically, once UTILISATION payment stops, so too does the ability to use the service.

> **UTILISATION**
> Two parties involved, payment for
> usage or time, ongoing

4. AFFILIATION *revenue model*

The AFFILIATION revenue model is similar to UTILISATION in that it's a 'rental', but this time there are three parties involved. This revenue model extracts payment for facilitating connection between two (or more) parties.

For example, LinkedIn charges an ongoing **membership fee** over and above their free service for those who want to use their platform in a more business-oriented manner. In this way, people are paying for more and better use of the platform to facilitate their AFFILIATION.

Examples of organisations that use this revenue model are:

» professional networking sites, such as Business Networking International (BNI) and Business Chicks

» dating sites, such as eHarmony

» membership networks, such as CPA, The Executive Connection (TEC), CEO Institute.

AFFILIATION
Three parties involved, payment
for usage or time, ongoing

It's important to note that just because a revenue model changes doesn't mean anything else in the Xcelerate framework has to change. We can change our revenue model without changing our business model, and vice versa.

A revolution for the price of evolution

The cool thing about revenue models is that they are:

» effective immediately, so you can impact your financials then and there

» easier to change than a business model

» simple to test and experiment with smaller sectors of your market

» able to offer your clients more choice in how they pay you

» able to be engineered to your customers' interests.

Sure, changing your revenue model still takes work and there's risk involved, but really, we're talking a revolution for the cost of an evolution.

Giving it air time

Jet aircraft engines were once manufactured and then sold for something like US$15–35 million. That meant a lot of powerful negotiating for the manufacturers – including Pratt and Whitney, Rolls Royce and GE – who were always getting their margins squeezed.

Competition was fierce. The two biggest buyers were Boeing and Airbus (think duopoly buying power like Aussie supermarket chains Coles and Woolworths). Scant profit was made on sale of the engines. Profit came from the maintenance and service agreements for post-sale work.

From the customer's perspective, it was difficult too. They had high fixed costs that they needed to budget for, their purchases were large and irregular and they needed to show a return on assets to shareholders. Once they'd decided which engine to buy, there was little recourse if it turned out it wasn't the best choice.

As argued by Dr. Henry Chesbrough, professor and Executive Director of the Center for Open Innovation at the Haas School of Business, over the years Rolls Royce coined, trademarked and pioneered the term 'Power by the Hour', which went some way to providing certainty to airlines on

service costs and incentivised the manufacturer to optimise airline uptime.

*But it was GE – a company that started with the humble light globe – that really revolutionised this space. They changed their revenue model from **TRANSACTION** to **UTILISATION** and started selling their engines for a fee per operating hour. (GE consolidated a number of their aviation service businesses under their OnPoint™ brand in 2005, and in 2016 they launched the True Choice™ product suite for their aviation services, which included True Choice™ Flight Hour services.)*

The customers were stoked!

GE had removed a major outlay and transformed a fixed cost to a variable cost based on performance. When the aircraft was on the ground for engine maintenance, neither company made money.

These are beautifully aligned incentives.

A revolution for the price of an evolution.

Brothers and sisters who don't fight

Business models and revenue models are like brothers and sisters who don't fight.

The right pairs are often found together, like the ones shown in Figure 6.2.

Figure 6.2: Business and revenue model combos

Business model	Revenue model	Example
PHYSICAL: *creation*	TRANSACTION	Adidas makes and sells shoes at a margin
FINANCIAL: *connection*	INTRODUCTION	A mortgage broker sets up the best loan for the purchaser for a commission
DIGITAL: *distribution*	UTILISATION	Spotify and Netflix allow music and movie streaming, respectively, for a monthly fee
KNOWLEDGE: *contracting*	TRANSACTION	A trade business marks up materials and labour for a job
KNOWLEDGE: *contracting*	UTILISATION	Professional services (e.g. a lawyer) charge a retainer
MARKETPLACE: *connection*	INTRODUCTION	eBay charges a fee for listing and commission on sales
SYNDICATE: *connection*	AFFILIATION	LinkedIn charges a monthly fee for premium membership

It is common for business models with the activity of **connection** (think brokers) to have a commission or a fee-based revenue model for putting two parties in contact. Consider real estate agents, mortgage brokers, insurance brokers, recruiters and online companies such as SEEK, Webjet and iSelect.

Whenever we see the business model activity **connection**, we find either the INTRODUCTION or AFFILIATION revenue model – since by default, a third party must be involved.

Look for areas where everyone has the same revenue model. This may be an opportunity to Xcelerate by changing quadrants.

Why rentals are a big move

More and more companies are moving their revenue models to UTILISATION or AFFILIATION, where 'rental' is involved.

The software industry, for example, has moved from selling licences (TRANSACTION) to selling subscriptions (UTILISATION), facilitated by advances in technology. Why?

Upgrades and newer versions are automatic. We often get extra devices for free. We don't need to worry about where we put the software discs when we buy a new device.

Look at movies and TV.

We watch our favourite shows whenever we want, for as long as we want, on services such as Netflix.

The switch to UTILISATION from TRANSACTION can be achieved without changing your business model. We all know we are paying more on monthly subscriptions than we used to pay for one-off TRANSACTION fees.

The shift of businesses in the developed world to become more service-oriented also predisposes companies toward employing a UTILISATION revenue model.

Just go to the gym

After one of my workshops, one of my clients was considering moving part of their offering to a UTILISATION model. I encouraged them to do some homework on what that might look like, calculate the costs, and then go and iterate their ideas with a couple of customers they worked closely with.

This client was a large provider of electrical contracting that had previously always charged its customers on a TRANSACTION basis. In thinking about moving to a UTILISATION model, the client came up with the following two lists.

Their customers' interests were:

» stability of spending, ensuring budgets weren't exceeded in individual months

» dedicated resources that grew to recognise their specific preferences

» more responsive service
(someone on-site all the time)

» fewer maintenance staff on their books

» fewer invoices to process, with less account payable time at their end

» more proactive and preventative work during quieter times

» agreed reporting structures and monitoring.

My clients' interests were:

- » managing the right staff levels to ensure service levels and response times with clearer requirements

- » the ability to plan and invest appropriately in labour and equipment for business needs with a clearer forward plan

- » customer-centered training and induction programs

- » stable income and profitability, rather than 'feast and famine'

- » reporting and review meetings to get closer to their clients' needs.

Clearly, this situation was win/win.

UTILISATION models often get a bad rap, because people feel that they are paying when they don't need to, or that they are paying whether or not they fully benefit from the product or service. For example, if I have a gym membership with a monthly subscription and I don't go, I lose and the gym wins.

But does the gym really win?

Sooner or later, I will stop going. The gym would much prefer I stayed and got my money's worth.

So if you want to change your revenue model, first ask yourself: **What's in it for your customer?**

Google it

Google introduced paid advertising in January 2000. The industry's ad model at that time was to sell ads based on cost per thousand impressions (CPM). Initially Google's ads were not 'self-serve'. In October of the same year Google decided to sell these ads online by credit card. This first edition of Google AdWords, still on a CPM basis, did much better. But it wasn't until Feb 2002, when Google launched a major overhaul of AdWords , that things really took off.

Prior to this overhaul, ads were being auctioned based on how much advertisers were willing to pay per click. This enabled anyone with money to outbid others, even with poor or irrelevant ads. Indeed, larger companies could easily out-bid smaller ones. Google recognised this was far from being in anyone's interests, having ads that were not clicked on where no-one made money. For the advertising to work, Google needed to highlight the ads that would be clicked on.

Wanting to reward the quality and relevance of an ad, Google introduced clickthrough rate. The algorithm that Google created to determine which ads are displayed where and in which order was based not only on the auctioned cost-per-click but also the clickthrough rate.

This gave searchers better and more relevant advertising, and ensured the overall system was self-improving. Introducing a quality variable was a great innovation to a UTILISATION revenue model (payment for clicks which 'rents' eyeballs). The result?

Google's overall revenue, still completely dominated by ad revenues, will easily exceed US$80B in 2016, up from US$74.5B in 2015.

Successful revenue model moves

As we've already said, revenue models are generally set up, like business models, at the start. But unlike a business model, they're usually easier to alter – not always, but certainly more often.

So let's look at some examples.

News Corp, which owns *The Australian* newspaper, have reasonably successfully transitioned some of their revenue to online subscriptions in addition to their traditional newspaper sales and advertising revenue. Their advertising revenue is UTILISATION (rental of space).

What about movies?

We used to rent them from places like Blockbuster (UTILISATION), then they went on to sell them in-store (TRANSACTION), then they went out of business because we started downloading them (TRANSACTION). Now we subscribe to services like Foxtel (UTILISATION).

And the future? Likely on-demand (for very little outlay).

Apple is a great example of how a business generates money from music sales (downloads: TRANSACTION), but has also developed a revenue model streaming music, which is essentially renting (UTILISATION).

We used to borrow books from a library and pay a membership fee (UTILISATION), or alternatively, we purchased books from a bookstore (TRANSACTION). Now we can buy books online from Amazon (TRANSACTION) or join the Kindle Owners Lending Library and pay an annual membership for a book per month (UTILISATION), which is just like a virtual library.

Costco combines a membership fee (UTILISATION) with purchase of products (TRANSACTION). It actually chose to have a hybrid revenue model in which an annual membership fee is also charged: a type of revenue model often associated with SYNDICATES.

Banks pay mortgage brokers a commission for each loan they facilitate (INTRODUCTION). Ditto for insurance brokers and real-estate agents.

Value is in the eye of the beholder

It's important to note that revenue models operate independently of pricing choices.

When prices are set for a subscription, for example, they can be made premium or cheap. They can be presented as three different choices for three different levels of access and they can be changed without changing the revenue model.

Trade-offs between price and volume and how exactly to present the offer can all happen within one revenue model.

Some companies can charge different amounts to different market segments (adult, student, concession, senior citizen). Other companies operate off list prices, specials, deep discounts, etc.

How you receive the money is not the same as how much you receive!

Revenue model wrap-up

Here's a list of some of the common pricing methodologies used within each of the four types of revenue model.

TRANSACTION

- » *make a margin*
- » *add a mark-up*
- » *discount off list price*
- » *specials*

UTILISATION

- » *licensing*
- » *franchising*
- » *loyalty programs*
- » *subscription*
- » *freemium*
- » *unlimited*
- » *per unit*
- » *per time*

INTRODUCTION

- » *commission, percentage of sale*
- » *commission, fee*
- » *commission, capped, uncapped, clawback, tiered*

AFFILIATION

- » *membership fee*
- » *subscription*
- » *access fee*
- » *advertising*

This is where Uber stands out. Unlike other **connectors**, who charge a commission on a price set by the seller, as Airbnb does, they chose a margin model within the INTRODUCTION revenue model.

Uber controls the market price and the wholesale price. Uber receives all the money from the passenger and then Uber pays the driver.

But guess what? This changes during peak periods and so do their margin dollars.

Let's Xcelerate Uber

There's lots of economic theory behind why Uber's model should work well in a supply-and-demand world. They call it 'surge pricing'.

It's not popular among those paying the higher price in peak periods (obviously), but the profitability of it is enormous and optimised in a way that is very rarely seen in other companies. (Uber markets this as a better way to ensure that drivers enter the market to drive when surge pricing kicks in. It ensures more supply is brought online by offering drivers higher prices to draw them in. It works pretty well, with most surge pricing periods being quite short).

Uber is doing well, but it would still be relatively easy for them to Xcelerate themselves.

Look at these two alternatives for Uber:

Alternative A

Uber could introduce a new offering that operated like a premium membership. They could use an

INTRODUCTION revenue model to launch an added service: e.g. an annual membership fee to get preferential treatment or additional services as a passenger, enabling car/driver preferences, or priority access in busy periods.

Alternative B

Uber could do some big data analysis of its client base and their spending or travelling patterns, then introduce a monthly subscription: e.g. the first three rides within 30km of the CBD for free.

Whatever, whenever, wherever

As we move more and more along the continuum of tailored 'for me' and 'on demand' – that is, what I want, when I want it, from wherever in the world I am – we can assume there will be more and more UTILISATION models that charge based on our own individual, specific usage.

This means payment based on our numbers of hours used, number of clicks, or number of transactions processed.

Your organisation must be sophisticated in its use of technology and big data to analyse client behaviour more deeply.

With time, your business will get better at working out the profitability of different customers and customer segments, becoming more capable of tailoring offerings for individual customers.

This will maximise the market size for your business as it optimises the cost of your product or service to its exact value for that customer. Optimising revenue and cost will thus maximise profit.

This will result in specific offers to individual consumers of products and services based on their unique requirements. Competition will increasingly strip away the fat until the end user is better off.

Economists call this first-degree price discrimination.

I suspect it won't be long before we have standing orders set up for $200 of groceries delivered per week. The shopping list will be populated by our past buying behaviours. The list will optimise our shopping baskets to our preferences for brands versus private labels and will act like a monthly budget for us.

Groceries will be delivered to us automatically each week and we'll be given the option to sign off or have it auto-filled. The service will be programmed to deliver 5–10% of new products, and will probably consider seasonality for fruit and what's in stock.

In this way, returns and waste at store level will be minimised. Hallelujah!

It's now up to you to determine how you might move your revenue model to embrace this opportunity and offer whatever your customer wants, whenever they want it, wherever they are.

Once you've mastered this, you can start thinking about how you communicate to your customers through the next model we'll explore: the communication model.

Xcelerate now

1. A revenue model answers these two questions:

    i. Will the customer **own** or **rent** your **ASSET**?

    ii. Are there **three** (or more) parties involved or just **two**?

2. A revenue model is easier to change than a business model.

3. You can change your revenue model without changing your business model and vice versa.

4. Revenue models can give you revolution (in performance) for the price of evolution.

5. Revenue models from other markets may be suitable for application in yours.

**All you
can fly**

7

Paying $400 for an airline seat is not unusual these days, but when Abram C. Pheil paid this price over 100 years ago, he was the first person to do that – ever.

Pheil was the first paying passenger on the world's first commercial flight.

He won the seat in an auction by a clear margin. It was the *only* passenger seat on the flight, which flew from St Petersburg to Tampa in the US with just the pilot and one passenger.

It had been ten years since the Wright brothers had made the first successful powered flight. Whilst $400 was a substantial sum back then, Pheil wasn't born into money. The article 'Abram C. Pheil' on the *earlyaviators* site goes into detail about how he'd moved to St Petersburg hearing of opportunities in 1894, landing a job in a sawmill at $1/day. He saved what he could and got a loan to purchase some plots of land, which he then built houses on. He managed to sell these for a profit, powering his future.

Many years later, his business acumen and drive enabled him to purchase the sawmill. Using sawdust from the mill, he took to filling in the ruts in the main street of St Petersburg in the days before it was paved. He continued to invest in property and businesses, purchasing the St Petersburg Novelty Works and arranging construction of two buildings in the main street. He was eventually elected Mayor of St Petersburg, and it was shortly after his term as mayor that he bid for his ticket on the first commercial flight.

Tony Jannus piloted the aircraft (and Pheil) across the bay at low altitude for 23 minutes, successfully completing that historic flight. Countless years later the field of aviation has seen many business models come into use, but interestingly it is in the area of revenue models that there have been some significant developments.

In 2013, Surf Air launched as an 'all-you-can-fly' airline. For a US$1000 sign-up fee and US$1,950 per month you get unlimited flights. You can check this out on their website **surfair.com**. Surf Air originally started with flights from Silicon Valley and Los Angeles operating out of the smaller airports of San Carlos and Burbank.

The idea is that for busy executives, the most precious asset is time. Moving in and out of smaller airports saves substantial time. Book in 30 seconds, board in 15 minutes. Add to that the ability to network with seven or eight other execs while aboard an airline dedicated to service and passenger knowledge, and you have a recipe for success.

Surf Air has 12 planes in use and 15 on order. This 'all-you-can' model is an example of the UTILISATION revenue model, and when applied to the airline industry it seems quite the innovative breakthrough. This revenue model

of 'all-you-can' is far from new, though, with all-you-can-eat restaurants and all-you-can-use gyms having been operating for years.

Following Surf Air, there have been a few additional start-ups launching into the aviation space with a UTILISATION revenue model.

In March 2016, OneGo started offering unlimited flying from 76 airports on 700-plus commercial airline routes in the US for a joining fee of US$495 plus a US$2950 monthly subscription.

And in Australia, our very own Airly is raising funds to offer travel between Melbourne (Essendon) and Sydney (Bankstown), just like Surf Air.

Surf Air, OneGo and Airly all provide a member-based subscription service for airline travellers. These companies share the same revenue model, but they have different business models.

I'll say that again. These companies have different business models, but the same revenue model. Call me crazy, but I think whether the companies own, charter or just book the planes may just be a tad relevant as an indication of differences in business model. Surf Air owns its own planes; OneGo utilises commercial passenger airlines; and Airly will use charter partners.

The type of service, and whether the planes are staffed with company employees or airline employees, is going to be important, as are the routes they offer and the airports they include.

These companies are great examples of the power of the revenue model. They show how easy it can be for an

incumbent to launch a company like this, instead of leaving the space empty for a disruptor to launch into this gap. Maybe they didn't want to; maybe they didn't like the economics; maybe they thought it would cannibalise their business; or maybe they just didn't think of it. This won't happen to you if you use the Xcelerate framework.

Now, remember that technology is an enabler and not a business model. Many people (not you, of course) would describe the companies we have mentioned as being disruptive tech companies. Well, don't Qantas and Webjet use the same technology? It isn't their use of the internet or an app that makes them disruptive – it's the thinking they did about how to use a different revenue model.

As we pull apart the business models and revenue models of anything shiny and new, we start to see some patterns. Then we start to see ways to take these evolutions, disruptions or leaps into other industries – or indeed into our own business worlds – to determine options for what to do next before anyone else does.

In the words of Aussie music critic Molly Meldrum, do yourself (and your business) a favour! Grab a model, locate yourself on it, then look around for other spots on the model where you can innovate the Xcelerate framework. You just might get revolution for the price of evolution.

Here are a few ideas that might fly:

» Maybe airlines could ride-share? Or conversely, should Uber introduce a subscription model for unlimited rides? (Like Airly)

» Should gyms try not owning their premises and instead just 'charter' their spaces? (Like OneGo or

charter airlines. EFM gyms have done this in schools and hospitals.)

» Can you consolidate purchasing power to get better discounts and then package that up in a subscription? (Like OneGo.)

» Can you use a new service to facilitate networking for business whilst people get what they need to do done?

» What if you give people access to unlimited usage of a service in return for a monthly subscription?

» How can you save busy people time? Because they will pay for that, for sure.

Communication model

8

When Atlassian listed on the NASDAQ stock exchange on 10 December 2015, it was the biggest-ever float of an Australian company. With the initial IPO valued at around $6 billion and oversubscribed, it ended its first day of trade up 32% at just shy of $8 billion.

Founded in 2002, Atlassian represented a long, successful and inspirational journey for co-founders Scott Farquhar and Mike Cannon-Brookes, who had met at the University of New South Wales. Their start was financed by a $10k credit card and remained bootstrapped for eight years.

Atlassian's products are software development tools that help developers collaborate. Two major products are JIRA and Confluence. The co-founders are now listed among Australia's top 20 richest, but they remain grounded.

One of the company values listed on their website is *'Don't Fuck the Customer'*. This tells us a fair bit about the culture of the company. Don't let the stories of Friday night poker and sponsoring beer at conferences fool you into thinking this

was not a company built by a couple of strategic dudes. Sure, they say some of it was fortuitous, but we ain't buying it.

Much of the early venture capital they received backed them personally as much as it backed their products. From early on, they debated their pricing and sales policy. They are now quite famous for turning the selling mores of business software sales upside down. They had no sales people, no sales team. This is rare in their space.

They facilitated this model by making everything transparent, such as pricing on their website. Everything customers needed to know and download and use their products was online. They wanted their product to sell itself. They were constantly receiving advice from everyone and everywhere that this model wasn't sustainable. As Cannon-Brookes says in the *Sydney Morning Herald* (2015), 'Almost everyone we spoke to would say well done, but you can't sustain that and grow without salespeople.'

So now at an $8 billion valuation with $457 million in annual sales, they have every right to declare the naysayers totally wrong.

As we have seen many times in this book, when the way a company operates is different to the prevailing model of an industry, the company can actually generate very powerful growth.

———————

Atlassian used a different communication model from many business software organisations that maintain a strong sales force presence.

How can I get more sales?

Revenue from customers is a company's lifeblood. When a company is looking to grow or address declining revenue, they often consider adding to or changing the money and resources they are deploying in communicating their message to prospects and customers. To say that most companies wrestle with this issue frequently is an understatement.

Have you ever asked or been asked these questions?

» Should we have a better digital strategy?

» How can we leverage social media better in our business?

» What place does content marketing have in our business?

» How effective is our current marketing spend? What's working best?

» Has advertising changed? Do we need to change where and how we advertise?

» How effective is my sales force?

» Does my sales team need more training?

» Should my team spend more time networking?

» Should I expand my marketing spend? Would that deliver additional results?

» Should I add more salespeople?

» Should we use distributors?

To answer these questions, you need a model that helps you think about how you are currently communicating to your market and to prompt new thinking about how to improve this.

This is where the communication model, shown in Figure 8.1, can help.

Figure 8.1: Communication model

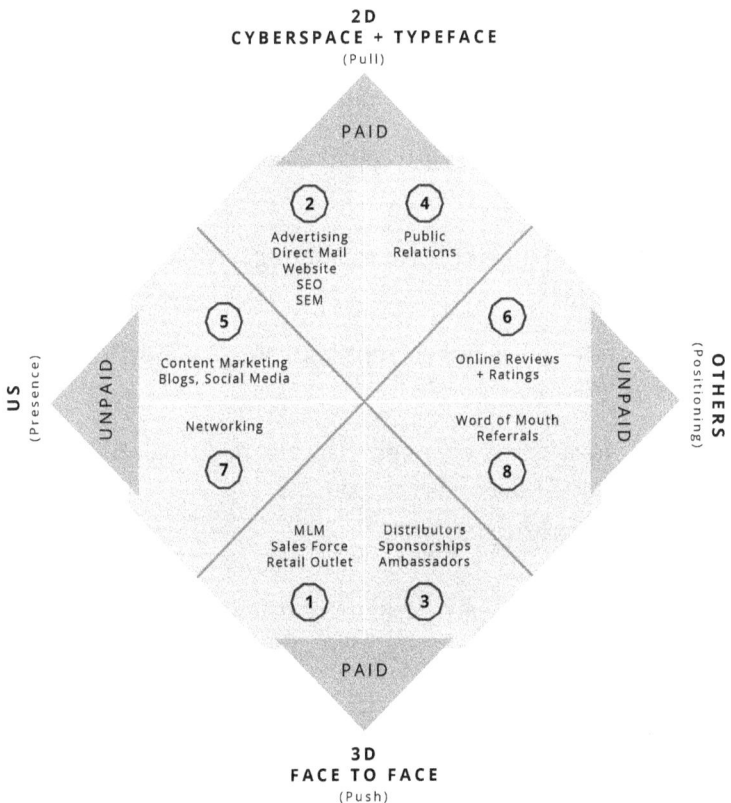

Banking on eight communication channels

Let's unpack Figure 8.1, which shows eight different ways, or channels, of communicating to our current and future customers.

The horizontal axis in Figure 8.1 indicates whether the message is coming from **US** directly or via **OTHERS**. When we deliver our message ourselves, it is best to have **Presence:** that is, humble authority, belief and conviction. (In the book *Conviction* by Peter Cook, Michael Henderson and Matt Church, they argue successfully that conviction is one of the key ingredients in successful selling.) When we have others delivering our message for us, we have **Positioning:** that is, we want our organisations to be known for owning a distinct place amidst a competitive landscape.

The vertical axis indicates whether the message is delivered **Face to Face** or via online, phone or print media, which we'll call **Cyberspace** and **Typeface**. This is basically anything that isn't a face-to-face meeting with a real live person. Another way to think about that is to label these 3D or 2D.

Atlassian eschewed Channel 1 and went for much more of a viral marketing approach that combined many of the other channels, but essentially their differentiation drove lots of **Word of Mouth**.

One of the big factors in determining your communication model mix is working out what you are willing and capable of **paying for** and what you won't pay for. You can see from the 'paid' and 'unpaid' labels in the model where things fall.

In the case of Atlassian, one of their early decisions about a sales force was easy; they couldn't afford any salespeople.

But later, when they could afford this, they had the clarity of strategy to hold their position – that they didn't need salespeople to continue to grow.

Companies often lack clarity and integration around their communication model. Many will work to spend in all areas, never really prioritising or measuring impacts.

Do we see Australia's 'Big Four' banks as being different from each other? Do we judge them on their TV ads or how they serve us day to day? How passionate are we for one bank over another? Do we switch as a result of their marketing? How much are they spending on marketing each year, would you guess?

These are really important questions to answer, especially if you are a bank or any other enterprise tipping massive sums into marketing when your customer base sees very little differentiation. Changing the emphasis of your communication model is something no doubt you wrestle with often, but what have you done to change the game lately?

Some examples of companies that have a fair amount of focus in the various channels are shown in Figure 8.2.

Figure 8.2: The eight communication channels

Channel	Description	Company example
1	Multi-level marketing (MLM) Sales force Retail outlets	Tupperware All real estate companies Telcos
2	Advertising – print Advertising – TV Advertising – social media Advertising – web (Google AdWords) Direct mail Websites and SEO	Banks Holden, Ford Groupon, Zynga Expedia, Priceline Wine sellers Carsales.com
3	Ambassadors – sports Ambassadors – perfume Ambassadors – watches Sponsorships Distributors	Nike Chanel Omega Bendigo and Adelaide Bank Sony
4	Public relations	Redbull & Apple
5	Content marketing	HubSpot
6	Online reviews and ratings	eBay Amazon Trivago
7	Networking	Business Network International
8	Word of mouth	Netflix Atlassian

Buyer behaviour
is changing

IF buying is changing, it follows that we may well need to alter the way we communicate. And buying **IS** certainly changing.

Over the years many businesses have had choices in how to spend their dough to increase sales. They could advertise to get their name out there (***Cyberspace and Typeface***) and build a brand, or they could employ a sales force (***Face to Face***) to carry their message to their existing and potential customers. Or they could do both.

One of the key differences between these two modes of promotion is that advertising and other forms of marketing promotion are typically one-way communication. If we watch an advert on TV or see one in a newspaper, the buyer can't have a conversation with the seller.

On the other hand, when a salesperson delivers the communication, there is a two-way interaction. Questions can be posed and answered. Offers can be tailored. Feedback can be taken back to HQ for debriefing and acting on.

The second difference between these two is that marketing communication can reach a lot more people more quickly than a sales force can. Given these differences, promotion and advertising is a bit like a sweeping broadsword while a crack sales team may be more like a sharp knife, depending on the business.

But buying is changing these pull and push forms of communication in the following ways:

» Print media and TV are being transformed. We're reading fewer newspapers and we're seeing many different viewing options away from free-to-air, such as pay TV, streaming options, social media and You Tube.

» Professional buyers and consumers alike are spending much more time online researching their potential purchasing decisions. CEB, a best practice insight and technology company, reports that the average purchase decision is 57% complete before a customer even engages with a supplier, and further that most buyers have consulted more than 10 information sources by this point.

» Technology is enabling better and better interaction online between buyers and sellers – think reviews and star ratings on Amazon and eBay, pop-up chat boxes, online forums, and soon, voice recorded questions.

» Social media enables better customer targeting – think age ranges, geographies, genders, preferences, and click history.

Suddenly, the one-way communication broadsword of marketing looks less attractive. The traditional media it is based upon is being eroded, plus the old way is *all one-way communication.*

And your sales force?

CEB reminds us frequently that sales forces are getting involved later and later in a buyer's decision-making process,

as the statistic above highlights. The average B2B decision-making group has on average 5.4 buyers and buying cycles are getting longer. Furthermore, CEB reports that 53% of customer loyalty is driven by the salesperson's ability to provide unique insights.

Our online world combines the 'reach' of marketing promotion with the 'trust' of two-way communication.

_____

Key questions to ask yourself

Some of the key things to consider when thinking about which communication approach to focus on, or which combination of approaches to use, include:

» *How big is my market? Can it be reached by direct sales? Could this be done cost-effectively? Atlassian said 'No way' to this. They were based in Sydney and their market was global.*

» *Is my product/service complex enough to require interaction for customers to choose the right offering?*

» *What is the value of my product/service, and how many sales pay for a salesperson? How does that fit with the market opportunity?*

» *Is the primary relationship with the product OR the company and their representatives?*

» *Am I facilitating a meeting place for others to interact? In that case, how do I need to communicate my message?*

» *Is the purchase process for my product/service moving more online? What am I doing about that?*

» *If I can't reach my target market effectively, would distributors, licensing or franchising work?*

Sales versus marketing

Marketing means many different things to many different people. Often we just mean promotion.

To me, *marketing* is about getting inquiries, leads and opportunities to have a sales conversation. It targets high numbers of people in proportion to those that actually inquire (often under 5%), and it's getting harder to get cut-through.

Selling is the science of moving people to a decision to purchase and helping them to act on that decision **at the time they have made it.**

So let's say:

Marketing = generation of inquiry

Sales   = generation of purchase decisions and action

Some companies are ***marketing-heavy*** (at the top of the communication model: **PULL**). For example: branded goods, telcos, banks, food manufacturers, cosmetics and supermarkets. *This is typical of B2C businesses.*

Some companies are ***sales-heavy*** (at the bottom of the communication model: **PUSH**). For example: services businesses, businesses with high-cost products (as opposed to consumer products), and businesses with products passing through distribution channels. *This is typical of B2B businesses.*

Some are both – for example, pharmaceutical or car industry companies.

SALES (Channel 1 on the communication model)

Best used when:

- » you know your potential customers and can get their attention cost-effectively

- » your message has some complexity and there is a buyer education aspect to it

- » it's more difficult to generate switching from competition or generate decisions

- » you distribute someone else's product

- » your offer is tailored and negotiated and the price varies

- » the relationship is more with the company/people and service is important.

MARKETING (Channel 2 on the communication model)

Best used when:

- » you want to build or maintain a brand

- » you want to reach a larger audience than you possibly could with a sales force

- » your target audience is more difficult to identify

- » your message is simple

- » you distribute your product to the market through others

- » the relationship is more with the brand/product, which plays the starring role.

What else do we know about sales and marketing?

In the B2B world, selling has far more effective conversion rates than marketing. In a piece of research conducted by Implisit, closed-won/closed-lost opportunity ratios were almost three times higher with sales-generated opportunities (57.5% versus 21.8%). Lead-to-deal rates were also higher for sales (0.94% versus 0.78%), although marketing (15.3%) performed better at lead-to-opportunity conversion than sales (13.9%).

Is direct sales dead?

A lot has been made of the death of direct sales. It's viewed by many as too old-school.

Three trends adding to this are:

1. the success and proliferation of the **MARKETPLACE :** *connection* and *distribution* business models

2. the easy availability of hitherto inaccessible information now via the internet

3. that online platforms and APPs are replacing traditional 2D communication forms and becoming hybrids that are more and more interactive.

Where does this leave anyone in sales? (Which is everyone interacting with a prospect or customer.)

The key shift that needs to occur, and it is very slow to happen, is that sales needs to become much more strategic, more insightful and make more links and connections to a

customer's specific challenges. It needs to be across all the technological advances changing the customer's business. Sales forces are striving to remain relevant to buyers when buyers can learn so much before they even speak to them.

This challenges sales to invest in their own development. The future salesperson needs to concern themself with two things:

1. How can I help my customer develop insights about their market?

2. How can I help my customer expand their market?

Both of these improvements will be welcomed by all buyers.

Personally, I don't think direct sales is going anywhere if it's used in the right circumstances, some of which we referenced earlier. But two things will change:

1. More purchasing will be online, so we will have fewer people working in sales.

2. Salespeople who endure in a shrinking pool will need to invest in their thinking to adapt to the changes we are seeing.

Ding dong!

David H. McConnell was in many ways an unlikely champion of empowerment for women. He was a struggling door-to-door salesman for Union Publishing Co. of Chicago, selling books. He wasn't earning enough money and wanted to try something to increase his success. So he started giving away free perfume samples as gifts to the homeowners he was calling on, in return for listening to his pitch.

He soon realised the perfume samples were in higher demand than the books he was selling. He also had a couple of other revelations. The householder that was home during the day (in the 1880s) was female. Few women were in the workforce in those draconian times. But more than that, McConnell realised they had a lot to offer the business world: more than they were given credit for.

He started the California Perfume Company in 1886 and began to hire women in rural areas who could tap into their networks of friends and family to sell perfume and other beauty products.

It was successful from the start. He enabled those who joined him to set up other sellers. And so began what is often now a controversial sales model called **Multi-Level Marketing**, **Network Marketing** *or* **Direct Selling**.

In the 1930s he launched the trademark Avon, which is how we still know the organisation today.

Famous for their 'Ding dong, Avon calling' jingle, they expanded to their peak US$11.1 billion annual revenue in 2012. However, their sales have been in decline for over four years now, down to US$6.1 billion. They have divested 80% of their American operations to private equity turnaround specialist Cerberus. US sales have declined to 14% of total revenue and more than halved to under US$1.4 billion since 2007.

Avon is now held up as an example of how the world is changing. Purchasing habits have changed, people argue. There's lots of other commentary too: Does becoming an 'Avon Lady' still appeal? The earnings aren't great and there are lots more opportunities to run businesses now. Attracting and retaining partners has been one of the biggest issues for Avon most recently.

*Multi-Level Marketing is much maligned and controversial. At its worst, it is viewed as pyramid selling. It is highlighted here as an example of a **Face to Face** channel communication model and the current relevance of this.*

There are in excess of 100 million people involved in Multi-Level Marketing according to Direct Selling

News, including Amway (1), Avon (2), Herbalife (3), Mary Kay (6) and Tupperware (9). Their performance varies by country with clear growth in emerging markets rather than from traditional large countries like US, Japan and Germany.

Is this an Avon issue or a communication model mix issue?

There are actually two issues at play. The direct selling model is being challenged by changing times and new competition. The selling method is being challenged and the appeal of the model for the sellers is being challenged. There are many other income-generating options that this needs to stack up against. And hallelujah for that.

Why you need a
Curtis Stone or Michael Jordan

In Channel 3 we are working to derive a benefit from a third party. The third party has something that appeals to us and/or our target audience. Done well, this channel can give more bang for the buck, but they do generally involve big bucks up front, so there is a fair bit of risk. Add to this the fact that we can't control the third party as easily as in more direct methods of communication.

THIRD PARTIES (Channel 3 on the communication model)

When might you want an Ambassador?

» you want people to identify your product/organisation with the Ambassador's positioning: e.g. to increase likeability, desirability, attractiveness

» you are building a brand and the Ambassador is a shortcut to elements of that brand

» you are reasonably unknown and haven't established your position in a prospect's mind

» the target market trusts the Ambassador more than your organisation.

When might you want a Distributor?

» you don't have a strong presence in some of your potential markets, which know the third party better

» you want blanket coverage, even up to saturation point

» you have a product, service, or system that is easily understood by the end customer

» you are large and bureaucratic to deal with

» it brings convenience, location, service, size, access, customer knowledge or bundling to the end user

» the distributor can offer ancillary and/or adjacent products or services that people prefer to buy together.

When might you want a Sponsorship?

» you believe the entity you are sponsoring will create cost-effective leverage for your message

» the sponsored entity matches the purpose of your organisation

» your target market is involved with the entity you are sponsoring

» there is a shared community

» it gives you access to an attractive network.

PUBLIC RELATIONS
(Channel 4 on the communication model)

There is a lot changing in PR as a result of everything we are talking about in this chapter, but also because it is an industry being disrupted by new technology trends. (For a great overview of the biggest challenges and opportunities facing PR agencies right now, Google *What Are the Biggest Challenges and Opportunities Facing PR Agencies now?* An article on Bulldog Reporter.)

Here are some of the changes:

» much more engagement in non-traditional media

» mobile content is paramount

» hyper-personalised content

- » the lines between internet marketing and PR are merging, as are the lines between ad agencies and PR. Organisations are looking for 'one story' for their audience.

- » measurement of impact is becoming more key

- » brands are looking for laser-focused targeting and personal connection

- » more noise and clutter to cut through

- » more tailored approaches, because each situation is different

- » staying up to date with technology and educating 'sometimes traditional' clients on the value of new approaches

- » real-time crisis management and dealing with bad press in whatever form and forum it comes

So why might you choose PR?

- » you want to get your message out fast and wide

- » you want their experience and network in how to do that

- » you want to build trust quickly by getting messages out through channels with more authority than you have

- » you want a savvy partner to strategically advise you in your communications

- » you want to manage risk and have solutions ready for negative situations

- » you want someone sitting over all your communications who gets you and your audience

What we won't pay for

SOCIAL MEDIA NETWORKS
(Channel 5 on the communication model)

Best used when:

- » you have an authentic, genuine approach

- » you are playing a bigger game than just sale of your product

- » you (and your audience) feel comfortable overlapping personal and professional boundaries

- » your company persona/brand resonates with personality (otherwise they can see you coming a mile away)

- » your message is narrow enough to generate a committed following

- » you have value to give prior to purchase and are willing to do so; thus, you are committed to content marketing and building your following.

I don't know about you, but most enterprises have a lot of trouble pulling this off. Larger companies are having a trust crisis at the moment as the rest of the world becomes more collaborative, creative and authentic. Traditionally they have just not been interesting enough and narrow enough (in audience terms) for their messages to be specific enough to resonate. They are also too big to relate to and usually don't come off as real.

ONLINE REVIEWS AND RATINGS
(Channel 6 on the communication model)

We mentioned earlier that one of the by-products of social media and online communities is that they enable two-way dialogue between sellers, buyers and others. More and more buyers are checking out reviews and ratings to make purchase decisions, so trust and interactivity are key to this. Socialbakers reports regularly on 'social care', suggesting that the best businesses in the world respond to nearly 100% of their social media service inquiries, and they do so within seven minutes to four hours.

This is an unpaid part of the communication model, so all influence is indirect. But there are some activities you can undertake to get improvements in ratings:

» operate transparently

» be open and deal with problems quickly and openly (Twitter receives more than 80% of service requests and only has a 20% response rate from companies. Facebook has a much better success rate of getting a response at 60%, but there are far fewer requests via this channel.)

» listen, be helpful and add more resources to your organisation to engage in dialogue with your customers and prospects

» make it easy for people to leave feedback

» review feedback daily and weekly

» develop guidelines for people to follow that are not phrased in corporate-speak, nor too bureaucratically long to be approved

» make measurements visible internally for all to see (and externally if you are going to be the leader).

NETWORKING
(Channel 7 on the communication model)

Best used when:

» both parties are targeting the same customer channels (e.g. professional services with 50+ employees who are tech enabled)

» their products complement yours and naturally go together

» you don't have access to the market yet, but your network does

» your organisations are of a similar size

» you are able to develop trust with the other parties

» you are at different stages of the same buying cycle (e.g. wedding venue, celebrant, florist, cake, band).

WORD OF MOUTH
(Channel 8 on the communication model)

This is the most elusive channel and deserves a bit more explaining.

Word of Mouth is by far the most powerful method of getting strong adoption of any kind. Businesses with great word of mouth need to do little else to communicate their message. Getting great word of mouth is easy to say, but hard to do.

When companies generate significant word of mouth, they rarely require the other channels of the communication model to be strong. In some of those cases, money still needs to be spent to build and impact positioning, but leads, inquiries and customers all come in significant numbers and are much easier to convert into clients. But companies only manage to achieve this in rare cases.

What are they saying?

Every September, Nielsen publishes a Global Trust in Advertising Report, which shows the percentages of global respondents who trust each advertising format, either completely or partially. Every year, Word of Mouth comes out on top.

In 2015, the top results were:

Recommendations from people I know	*83%*	*(Channel 8 – Word of mouth)*
Branded websites	*70%*	*(Channel 2 – Websites)*
Consumer opinions posted online	*66%*	*(Channel 6 – Online reviews and ratings)*
Editorial content (e.g. newspaper articles)	*66%*	*(Channel 4 – Public Relations)*
Brand sponsorships	*61%*	*(Channel 3 – Sponsorships)*
Emails I signed up for	*56%*	*(Channel 5 – Newsletter, blogs)*

Do successful companies spend significant money on sales and marketing?

Now that is a big question. The answer is no. Most large companies do spend a lot on sales and marketing – BUT many successful companies in their growth phases spend very little money on sales and marketing promotion.

You already know all the big examples, such as Amazon, Netflix and the celebrated Australian Atlassian. So how do we reconcile the fact that some of the most successful companies ever spent little money on sales and marketing to achieve substantial growth?

The answer lies in where their business comes from. It comes from **Word of Mouth**. When something is new and special and a company does a great job with their differentiation, people talk about them. When they do an amazing job, people rave about them.

Many of today's 'unicorns' and newer listed companies have grown to hundreds of millions of dollars in revenue without much traditional sales and marketing.

The secret of their success?

The answer lies in the differentiation model, which we will cover next.

Xcelerate now

1. The communication model comprises eight channels for communicating your offering to prospects and customers. The channels involve the following four perspectives:

   i. Presence and Positioning

   ii. Paid and Unpaid

   iii. Push versus Pull

   iv. 2D and 3D.

2. Marketing is leads; sales is conversion. Marketing has reach; sales creates action.

3. The obligations of social media communication for organisations are becoming much more demanding (in terms of percentage of queries responded to and time frames in which to do so).

4. Buying behaviour is changing. Much research is completed before the buyer even engages, so we need to adapt how we communicate to buyers.

5. Word of mouth is the most powerful communication channel.

Tinker, tailor, solder, spy

9

In Houston, Texas, a young lad of 12 years was collecting stamps and noticed that prices were beginning to increase. He set about producing a catalogue with 12 pages of his own and his friends' stamps. He advertised his catalogue in a stamp collectors' magazine and made US$2,000 by selling **direct** to the collectors. This was 1977 when the average annual salary was under US$10,000 for an adult.

His mum was a stockbroker, and he managed to invest some of this money in stocks and precious metals.

In early high school this business 'whiz in the making' was exposed to a data processor, which fascinated him. He spent time investigating and tinkering with computers at Radio Shack and at the age of 15 bought his first computer with his savings, an Apple II. As part of a sequence of events testing the bounds of his parents' patience, he set about pulling apart his brand-new computer to analyse how it worked.

Around this time, our young entrepreneur was making additional money selling subscriptions to the *Houston Post*.

He was cold-calling to do this, and it struck him there must be an easier and quicker way to sell. He hypothesised that the two groups of people most likely to purchase a subscription to the *Houston Post* were couples recently married and people who were moving house. So he researched lists of marriage license applicants and mortgage applicants. He wrote them personalised sales letters on his Apple II and sent them off. He made US$18,000 that year – more than his history teacher. This was no ordinary kid.

The following year he bought an IBM desktop computer. He taught himself how to upgrade and augment these with new components. He began to buy, repair and build IBM-compatible computers. He bought out-dated models from retailers and upgraded them and sold upgrade kits to others. He soon began sourcing parts direct from wholesalers. Most important of all, he sold his computers. He went from tinkering, to tailoring, to soldering to spying. **Tinker, Tailor, Solder, Spy.**

By the end of high school, all this activity was becoming a 'distraction' from his schoolwork. His parents had always wanted him to follow in his father's footsteps and become a physician. He enrolled in pre-med at the University of Texas in 1983 to appease them. Continuing his IBM-compatible upgrades and selling to friends, students and local businesspeople via word of mouth, he quickly outgrew his dorm room and moved into a condominium, resolving to drop out of pre-med. In a last-ditch effort, his parents – bordering on ropeable at the prospect of their son dropping out of college – put enough pressure on their wayward son that he agreed to go back to college if his sales over summer were disappointing. Sales in that first month were US$180,000 as reported in *Business Review Weelky*, Volume 26.

And so, following one of the fastest business start-ups in history, Michael Dell's business grew and grew and grew. In addition to his clear gift for electronics, Dell demonstrated extraordinary business savvy as well as the vision to orient back to north when bumped off course. No medical career for this dude. Sorry Mum and Dad.

During his 'Taking the direct approach' interview with *Entrepreneur Magazine*, Dell revealed that his first company PC's Limited sold US$6 million in its initial year of operation (1983). By 1986 sales were US$34 million. In 1987, PC's Limited became Dell Computer Corp. By the end of 1988 sales exceeded US$159 million; by 1991, US$800 million; and by 1992, over US$2 billion.

Dell did a lot right in the early days, but his insights were the most powerful, befitting one of the most impressive business visionaries of all time.

> **Insight 1** – Just like his stamps and his paper subscriptions, Dell sold *direct* to his customers, via phone, fax and mail. At that time IBM and Compaq sold through retail outlets. By selling *direct* to customers, Dell could sell at 15% below competitor's prices.

> **Insight 2** – Dell initially advertised to savvy computer users, who knew what they wanted. He placed advertisements in PC mags, local newspapers and even on the back of pizza boxes.

*If you look at the communication model, he shifted the industry norm of **Face-to-Face** (Channel 1) to **Typeface Advertising** (Channel 2).*

_____

Insight 3 – Dell made to order. Because he had more direct contact with the customers and would tailor computers to their needs, he saw what they were ordering, what they wanted. This tailoring component was new and different.

Insight 4 – The made-to-order model meant he kept much lower stock levels than his main competitors, had less waste and could supply the latest gear quickly with good supplier support. This further enabled him to keep costs lower on great quality computers.

*If you look at the communication model, he generated a lot of **buzz**. He got significant **Word of Mouth** – other people were selling for him. (Channel 8)*

_____

By 1996, Dell's sales were US$5.5 billion. He had successfully championed a new sales and communication model in the PC market, but he was not done yet. In 1996, while many of his competitors were still making their mind up about selling over the internet, he launched one of the first online websites for PC sales. Dell.com began selling US$2 million a day online. Between July 1996 and February 1997, sales increased to US$7.7 billion.

Reflecting on the Xcelerate framework, we see that Dell's business model didn't change – it was still **PHYSICAL: *creation***, the same as his competitors. His revenue model was still **TRANSACTION**.

But his communication model changed substantially from the industry norms of communicating and selling via retail stores and the salespeople in them (***Face-to-Face***) to advertising and direct selling via phone, fax and ultimately online (***Cyberspace and Typeface***).

When we cover the differentiation model, we'll see how he also drove word of mouth from that perspective.

Differentiation model

10

McDonald's provides a lot of companies with a lot of food for thought. Maccas, as we love to call them in Australia, are often held up as the epitome of operational excellence, with optimised systems and processes at the lowest cost. They have owned the low-cost burger segment since businessman Ray Kroc first joined the McDonald brothers' operation in 1954.

But here's where it gets interesting.

At McDonald's we can now:

  » 'create' our own burgers

  » watch our burgers being made

  » order electronically

  » get table service

  » eat burgers two or three times the height they used to be

» drive through and order coffee of *much* better quality than it used to be.

This is a non-standard, tailored burger with higher quality products, a variety of ingredient choices and personalised service. None of this says 'cheap burger' anymore.

And what happens when you increase choice and customer service?

Price increases or profits decrease.

This is a change in strategy from its heritage and it's a bold one. For decades McDonald's has meant cheap, standardised burgers. Now it's reaching for more. This change is on the back of changing consumer preferences and declining financial performance. It will be fascinating to see if it works.

Lessons from Ronald

McDonald's Australia is a leading example of the global McDonald's success. The first McCafé launched in Melbourne, Australia, in 1993. Around 300 were established in other parts of the world before the first McCafé opened in the US, eight years after Australia's first. The McCafé is said to have added 15% revenue to McDonald's restaurants that had one.

So far this has worked out well for McDonald's in Australia, which reports better growth results than its brothers and sisters in the US market.

If any company can pull off a major departure from their traditional market differentiation, then we'd expect it to be McDonald's. It has the resources and a great track record in Australia of getting its consumers right.

However, Maccas needs to be careful that it doesn't try to be all things to all burger eaters. At the end of the day, does it stand for cheap burgers? Quick and convenient burgers? Tasty burgers? Or tailored burgers?

This is a tough ambition to digest.

How does this explain differentiation?

Maccas always stood for cheap burgers. Hungry Jacks emphasised that 'the burgers taste better'. Grill'd went for being part of their local community, from produce sourced locally to giving back to local communities, aiming for an emotional connection to the brand.

The subject of differentiation – that is, how a company can grow, dominate, and win business using its 'difference' – doesn't seem to have advanced in the past few decades.

In 1980, Michael Porter introduced us to his 'Three Generic Strategies' to achieve above-average performance in any industry:

The three 'generic strategies' were:

1. cost leadership

2. differentiation (drivers defined very broadly as uniqueness in product and services, features and performance, personnel, technology and quality, etc.)

3. focus (as in 1. and 2. above, but focused in a narrow segment and tailored for the needs of buyers, aka customers, in that segment)

Thirteen years later, in 1993, Treacy and Wiersema wrote a *Harvard Business Review* article and a book on the three 'value disciplines':

They postulated there were three types of company, and that the more a company focused on *beating* the market in one category and *meeting* the market in the other two, the better its performance would be.

The three 'value disciplines' were

1. operational excellence

2. product innovation

3. customer intimacy

Then, in 1999, Hagel and Singer came up with three types of businesses in their *Harvard Businees Review* paper 'Unbundling the Corporation'. They argued that it was very difficult for these three business types to coexist, and that trying to be more than one would lead to significant trade-offs. Their call to action was to 'unbundle' to survive – that businesses should make a definitive decision about which type of business to focus on from the three below:

1. infrastructure (like low unit costs, economies of scale, high volume)

2. product innovation

3. customer relationship.

Can you see the pattern?

Each of these three examples, which span a period of two decades, presents essentially the *same* three ways to differentiate:

1. cost leadership

2. product innovation

3. customer focus (instead of focus, this might be similarly described as intimacy, relationship, service, experience, etc.)

The rule of three

To make things simpler, I like to show each of these three scenarios from the customer-benefit perspective. This is the external perception, rather than the internal orientation, as shown in Figure 10.1.

Figure 10.1: The three traditional ways companies have differentiated

Internal Company Goal (INPUT)	Customer View (OUTPUT)	Achieved by
Cost leadership	Lowest price	Operational excellence, systems, processes, technology to lower costs, economies of scale, high volumes
Product innovation	Best product(s)	Resources to innovate (time, high quality personnel, research and development budget)
Customer focus	Best customer service	Customer surveys and feedback, thorough staff training, clear service culture and values

From here forward, let's define the three ways you can differentiate by their customer benefit:

1. **Lowest price** – i.e. cheapest product. Think Aldi or Jetstar.

2. **Best product(s)** – i.e. best on the market in most people's eyes. Consumers will travel, wait, and/or ignore poor service to get these products. In the past it's been Apple product launches. Currently, it's Tesla.

3. **Best customer service** – i.e. always put the customer first. Tremendous choice is offered, company plans will be altered to suit customers and services will be tailored to meet a specific customer requirement. When Starbucks first entered the US market, you could get coffee any way you wanted it and a place to enjoy it. (I have even seen Starbucks staff sing to their customers on a cold, snowy day in Chicago.)

All three of the perspectives on differentiation (Porter, Treacy and Wiersema, Hagel and Singer) agree that not only should companies put one of these dimensions at the forefront of their approach, but that it is impossible to excel at all three.

Above all, companies should avoid getting 'stuck in the middle' and being average, beige, vanilla, mediocre and bland in each.

———————

This makes sense, right? If I am a product innovation-focused company, I need to hire bright minds and conduct a lot of research and development. It is very *unlikely* that I can offer the lowest cost as well.

Also, we tend to find that companies with great products have poorer service. (For example, restaurants with great views often have poor service because people will come for the views even when the service isn't great). You may even remember that in the early days of Apple iPhones, customer service was terrible.

Let's adapt Figure 10.1 and add in some additional company examples so we can see how they differentiate (Figure 10.2).

Figure 10.2: Companies and the way they differentiate

Internal Company Goal	Customer View	Australia	USA	Rest of the World
Cost leadership	Lowest price	Bunnings, Rio Tinto	Costco, Walmart	Toyota
Product innovation	Best product(s)	CSL	Apple (products)	Rolls Royce, Novartis
Customer focus	Best customer service	iiNet Athlete's Foot	Zappos, Ritz-Carlton, Nordstrom, Starbucks	*Can you think of any? It's pretty tricky. I'm out of ideas!*

Pick one and practice it

I've helped businesses with this concept for quite some time. When you focus on **one** of these three orientations, you can differentiate yourself from your competition in the eyes of your customers and achieve great results.

This **work** grows businesses. And it is work.

The trick with the differentiation model is that you should rarely change from one orientation to another – e.g. from a customer-focused organisation to a cost leadership one. That would be like Aldi changing to a customer service company or Apple having the cheapest products. (This is what McDonald's is trying to do: offer great burgers with low prices and fast, convenient service – having all three.)

No, instead you need to double down, go all in, and put your current orientation on steroids.

So if you are a customer-focused organisation, then you need to focus on getting better, continuously, now and forever.

Many struggle with the idea that to get stronger we need to narrow our focus, not broaden it.

We're all human. We experience FOMO (fear of missing out). We want to be everything to all people. Most of us worry that *if we choose one, we may lose some.*

If we focus on customer or product, then we will miss out on those who buy on price.

But if we try to be the best at all three, then we become mediocre in all three.

*The biggest secret in
successful business?
Go narrow to get bigger.
Narrow your business focus and
you'll get better at that choice.*

———————————

It all depends on your bedside manner, doctor

It's difficult for a **service-based** business to show It lias a **better product** than a competitor.

Professional services (accountants, lawyers, doctors), financial services and trade services businesses often run into challenges demonstrating that they have a leading product. An accounting firm, a legal practice, or an electrical contractor, for example, would find it difficult to demonstrate that its work is the best. It is seldom compared. How would we know if the work they did was the best or not?

What is compared is their customer service.

Customers can't really compare the 'product' in these businesses, so most of the time you are better off differentiating on customer service.

Think about your standard experience in the doctor's office.

It's hard to discern the skill of one GP from the next. Who is the best doctor? Who offers the best 'product'? Who can make you better fastest?

Doctors tend not to market themselves as 'the cheapest', unless they bulk bill. So they are really left with no choice but to differentiate on their bedside manner. If you find a doctor with a bedside manner you like, then you'll keep them for the rarity they are.

The rest are too busy maximising billing by whisking us in and out of the treatment room in record time. How often do you trot after them down a hallway?

If you're a doctor, then here's a big tip: be friendly and interested, spend more time with your patients, charge a bit more for the service and you'll be more successful. You'll enjoy your work more too.

Professional services and other **KNOWLEDGE:** *contracting* business models are ripe for disruption – and it's already begun with online doctors.

Price, product or service?

It's easier to identify an organisation that differentiates on price or product. Lowest price and best product are reasonably self-evident.

*So differentiate your business on **customer service** if it's difficult:*

» *for buyers to tell the difference between your organisation and the competition's*

» *to compare your product/service attributes*

» *to develop a leading product/service (that people will travel to or wait for)*

» *to retain customers (lots of choice, low switching costs, low loyalty).*

The Bain of my existence

There is a major disconnect between a company's perception of their customer service and their customer's perception of the service.

Not everyone can have the cheapest price, and not everyone has what it takes to continuously innovate products, which leaves many with just customer service to differentiate.

Why then is it so rare for organisations to generate difference from how they engage, service and woo customers?

If we take a look at our biggest institutions, it's very hard to tell which bank has the best products. Or the cheapest prices. Or the best service.

They all seem remarkably similar, right?

In 2013, consultancy Bain and Company published a research report that detailed NPS scores (Net Promoter Score is the percentage of 'promoters' minus the percentage of 'detractors') from interviews with 9,000 Australian consumers across 19 industry segments. The average NPS scores from 16 of the 19 industry sectors were negative, indicating there were more detractors than promoters.

Our largest organisations reflect that it's seriously difficult to train and empower, lead and facilitate their organisations to customer service glory.

» Is it because of the distance between management and customer? Too many layers?

» Is it complacency or delusion?

» Is it lack of awareness?

> » Is it an inability to shift the organisation meaningfully?

> » Or is the value in doing so just under-appreciated?

This issue of lack of service (you know, the hours wasted in phone queues each year for our largest banks, telcos, insurers, utilities and service providers) is the Bain of my existence.

What you can learn from Joshie the Giraffe

The luxury hotel chain Ritz-Carlton is renowned for its customer service. In fact, they have a Leadership Center dedicated to it.

Here's why they stand out:

Training – Traditionally, Ritz-Carlton employees have been trained in the use of language and expressions: what to say and what not to say, what to do and what not. This has created a collective identity and superior standard across the group of luxury hotels. Saying, 'My pleasure' instead of 'OK' is a simple example. But they also regularly survey their customers and act on this feedback via training.

Empowerment – Ritz-Carlton employees have a $2000 discretionary limit that they as individuals can choose to deploy against a customer problem. Taken from their website, here are the reasons empowerment is so important to them:

1. Employees are more accountable

2. Employees are more attentive

3. Employees will feel more valued

4. Employees will be more invested in work

5. Problems are resolved faster

6. Customers experience better service

7. Organisations are more nimble.

For an example of the lengths that Ritz-Carlton staff will go to, just Google 'Joshie the Giraffe'.

Then ask yourself: *Would my company do that?*

Breaking the rules

Focusing on any one of the three business dimensions leads to better performance, but some companies seem to bend the rules. Not the ones we see everyday, but certainly the Xcelerators.

If it's so difficult for companies to be all three types of organisation (and it is), then how can we explain some very successful companies who seem to be good at more than one?

When Amazon launched, they differentiated at the **lowest price**. But each year Amazon is consistently voted one of the best **customer service** companies. And whilst they continue to do both of these things, they keep coming up with new products like warehousing services, data centres, Amazon Web Services, Kindle, Echo and Amazon Go. So **product innovation** isn't a million miles from them either.

Netflix started mail order DVDs cheaper than Blockbuster,

but they tracked their users' preferences better than anyone, and then they started creating their own content. That's all three – again.

Dell aimed for 'cheaper' and they tailored to consumer preferences as well as providing great service. All three – this is a pattern.

Xero accounting software blasted into MYOB's scene in the SME space by innovating cloud software that was more user-friendly for small business, as well as great online training tutorials and live data feeds that saved bookkeepers time spent entering information.

Uber is cheaper, often preferred to incumbent taxi organisations and uses new technology to innovate the booking experience.

The fourth dimension

All of these companies have changed at least one model in the Xcelerator framework – business model, revenue model and/or communication model.

They have all changed the prevailing norms of the industries they entered.

They have all changed the rules.

It's as though the game got re-set, and the rule of not trying to be all three hasn't applied.

They have all continued to maintain their advantage over time. Not perfectly, not without some issues here and there, but quite impressively nonetheless.

But how?

My answer to this is that they maintained a *market innovation perspective*. They wanted to innovate the way business was done in their industry, and then they kept going.

Each time they innovated the *way they worked*, they made more money and they could afford to spread their resources across multiple dimensions. Often their Xcelerate framework innovations gave them a cost advantage, and often their changes to the way business was done were preferred by their customers too.

Basically, they were extraordinary Xcelerators.

This takes tremendous smarts, amazing vision and relentless drive.

Expand, shift, create your market

In Steve Blank's book The Four Steps to the Epiphany *(2003), he explains that a startup isn't just a smaller version of a larger incmbent organisation.*

*He argues that market incumbents have customers, know them well and are in the business of **product development.***

*Conversely, startups have no customers and need to work on **customer development**, all the while tailoring and adjusting their product according to feedback from early adopter customers. Makes sense, right?*

This is a powerful concept, and it's interesting to think about how large and small companies might not be great at imitating each other.

Put another way, the little kids have a lot to learn and lots of mistakes to make, but they tend to be a bit bolder and push boundaries, not knowing the dangers they will encounter.

The big kids have already made lots of mistakes and are more likely to be careful, avoid risk and play within their limits.

*You know intuitively
that enterprises are
not like startups.
Enterprises have
hierarchies, history
and hubris.*

———

When enterprises aren't working on product development, they're often concentrating on optimising, systemising, controlling and monitoring the company to increase productivity and profitability. This is the trademark of many larger organisations from the Industrial Era – in other words, **company development.**

As we've explored, there is another innovation dimension to all this.

In addition to product, company and customer development, there are now companies that are specialising in **market development**. Some of the most successful companies at the moment (Google,

Amazon, Netflix, Uber, Tesla) are innovating their markets by changing their Xcelerate framework.

Changing the way industries operate can **expand**, **shift** or **create** markets. When **markets** are innovated by changing the prevailing way business is done, we see companies defined far less by their products or customers and much more by their ability to move markets.

Now that Uber Eats and Uber Rush exist, you can no longer categorise Uber strictly in the taxi and ride-sharing game.

———

In the same way, we have already seen:

- *Amazon become the biggest B2B web services company in the world – it owns lots of bricks-and-mortar warehouses and at the same time has the biggest online B2C buying platform, not to mention their burgeoning consumer products business (Kindle, Echo, Alexa).*

- *Netflix has totally changed the way people pay for, receive and consume entertainment. Twice.*

- *Google is still the generic word for 'search' and yet they are most commonly cited in the news for their driverless cars. They also have Google Hangouts, Google Docs, Google Voice, Google Drive, Google Shopping, YouTube and Google Maps.*

Of course, companies should undertake all four kinds of development at different times and stages, but the extent to which they choose one dimension to focus on is the extent to which they are differentiated.

A new way of working

Here is a new differentiation model that better maps those observations, especially in periods of high change (such as industrial revolutions) when technology is enabling new business models.

Figure 10.3: Differentiation model

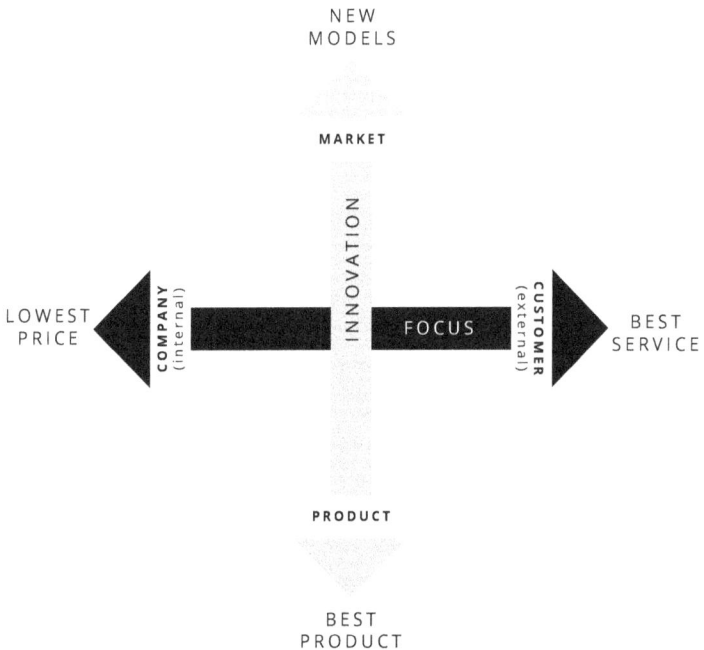

NEW
MODELS

MARKET

INNOVATION

LOWEST
PRICE

COMPANY
(internal)

FOCUS

CUSTOMER
(external)

BEST
SERVICE

PRODUCT

BEST
PRODUCT

The horizontal axis on Figure 10.3 shows what the business is focused on: INTERNAL or EXTERNAL improvement.

This is the same as **Cost Leadership** (INTERNAL) and **Customer Focus** (EXTERNAL).

The vertical axis shows whether the organisation is in the business of **product innovation** or **market innovation**.

Market innovation means organisations are changing their Xcelerate framework models to impact the way business is done in a market.

This provides a great counterpoint to the established rule of three and helps us to explain some of the exceptions.

More importantly, it helps you create even better businesses.

Snapshot summary

» **Market development** – changes the prevailing way business is done and is more about the strategy, not the weapon

» **Customer development** – focuses on developing early adopters for sales of new offerings, user experience, user feedback and customer field research

» **Product development** – innovates efforts focused on building a better mousetrap, a better weapon

» **Company development** – works on operational excellence, scale, efficiencies, optimising systems and processes

Standards versus brands

Standards and brands – are they special cases?

A standard is what Google has become to searching, eBay has become to trading and Facebook has become to connecting family and friends. Once so many people 'convert' to a format or way of interfacing, there are huge barriers to switching.

Each of these organisations arrived at their destination as standards via the user experiences that gained them mass followings.

Google provided the most applicable search results. Facebook worked better than Myspace. eBay was one of the earliest and most user-friendly buying and selling marketplaces.

Each company has the best product in its space.

Google has adopted fourth-dimension **market innovation,** having a vast offering beyond mere internet searches: for example, Google X's 'moon-shot' projects. Facebook is also showing signs of adopting the **market innovation** dimension. Standards are born of one of the four dimensions, and they can only be slain by companies that are good enough at differentiating as well.

In the context of our differentiation model, brands are good examples of taking customer service to an emotional and connection extreme (in the positioning of a brand).

Brands work to stay true to themselves whilst making sure they resonate with their target audience. The audience receives some emotional benefit, cachet, acceptance/ belonging, or identification with the brand.

Once brands are built, prices and products can arguably be secondary, but only up to a point.

An Amazonian mission – Go!

'There are two kinds of companies: those that work to try to charge more and those that work to charge less. We will be the second,' said Jeff Bezos, founder of Amazon.

'We've had three big ideas at Amazon that we've stuck with for 18 years, and they're the reason we're successful: Put the customer first. Invent. And be patient.'

Amazon's revenue has grown tenfold in the past nine years to US$107 billion (2015) and is on track for about US$135 billion for 2016.

The company aims for cost leadership, product innovation and customer focus, all while maintaining market innovation.

Jeff Bezos began with a big vision for something online. He also began his business from his garage in 1995. That was about the last thing Bezos did that was a cliché.

Amazon began by selling books online. This quickly expanded to CDs and DVDs (1998), then toys and electronics (1999) and ultimately shoes, jewellery,

*home improvement gear, software, gifts, video games and cameras. Selling stuff online was his gig. (**PHYSICAL: distribution***).*

*As Amazon's sales have grown, they have invested, somewhat ironically, in bricks-and-mortar properties in the form of many warehouses or 'fulfilment centres'. In 2000 Amazon launched their third-party seller business whereby companies could outsource the sales, storage, delivery, returns and customer service of their products. This now represents over 40% of their revenue and billions of dollars in sales. This business model is **PHYSICAL and KNOWLEDGE: contracting.***

*In 2002, Bezos launched Amazon Web Services (AWS), their cloud computing platform. AWS (**DIGITAL: creation**) offers developers a suite of services to add capacity and speed at more affordable rates than site-based servers. NASA, Netflix and the CIA are all major clients of AWS. The service was dubbed IaaS (Infrastructure as a service). As of 2015, Amazon had the cloud computing capacity of ten times the next 14 competitors combined.*

*In 2006, Amazon launched Amazon Fresh, its online fresh food delivery business in Seattle. It offers fresh food delivery the same or next day across a dozen or so cities on the east and west coasts of the USA, plus a couple of international sites (**PHYSICAL:***

distribution). They have now most recently launched Amazon Go (2016), applying this same business model to bricks-and-mortar supermarkets. This is a product innovation in the supermarket game.

2007 saw the introduction of the Kindle e-reader for book downloads. (**DIGITAL: distribution** for the books and **PHYSICAL: creation** for the Kindle itself)

In 2014, Amazon launched their Echo voice-activated information device. The Echo acts as a smart hub for a home's connectivity as well as playing music, making to-do lists and connecting to weather reports. (**PHYSICAL: creation**)

Among a number of significant acquisitions (including Zappos) was Kiva, a robotics company (2012), and the Washington Post (2015).

And let's not even talk about the planned drone delivery service ...

All this is not too shabby for a company that is just over 21 years old.

The last thought I'll leave you with on Amazon is that Bezos leaves an empty chair around the table in meetings to remind his managers that this is where the customer is sitting and that they must look after them.

The last word

In 8: Communication model, we talked about the power of **Word of Mouth**.

Intuitively, you know that having others talking about your products and services is far better than you talking about them. Whether customers do this or not depends on *the extent to which you have differentiated.*

I have never seen an advert for Uber, Netflix or Amazon. I am sure there are some, but that's not how they grew.

Most companies come up short of remarkable. It is really easy to try to be all things to all people, and really hard to have the discipline and conviction to narrow. Just like a sharper knife cuts better than a blunt one, the same force of a company's total resources applied over a narrower area gets better cut-through. The same principle applies to aligning your resources along a differentiation orientation.

Next, we'll look at one example of a company leading the way in differentiation.

Xcelerate now

1. The best business minds focus on one of three disciplines when it comes to differentiating – cost leadership, product innovation or customer focus. The result of these disciplines for customers is low price, best product or best customer service.

2. Companies that achieve focus in just one discipline outperform their competition.

3. Most companies try to excel at all three disciplines, and this makes them mediocre at all three.

4. When we have a technology revolution, we see the Xcelerate framework change more often, and this gives rise to a fourth dimension.

5. The fourth dimension is market innovation: new ways of working.

Wired for sound

11

Tell me a product that you can find in your car, your home, your carry-on on a business flight, in a NASA aircraft and an Olympic stadium.

You'll have to listen for the answer.

Amar G. Bose was born in Philadelphia in 1929 to an Indian father and American mother. As a student of physics at Calcutta University, his father had been a part of the Indian Independence movement, which landed him a short stint in prison before he emigrated (escaped) to the US.

By the time Amar was 13 he was adept at electronics, able to fix many appliances. Bose's father generated his income from importing coconut fibre mats from India, but his business came under great strain during WWII when all non-military shipping was ceased. Young Amar suggested his father put up a sign advertising radio repairs in the hardware stores where he had been selling his mats. And so they did.

Bose went on to leave an indelible mark in several fields, but he is most famous for founding Bose Corporation in 1964,

which revolutionised audio technology. With his early passion for music and an ear for sound from playing the violin, he dedicated his life to the relentless research, invention and commercialisation of products that improved what we hear. It is undeniable that Bose Corporation became a **product innovation** company of exceptional quality.

Bose attended the prestigious MIT to earn his Bachelor of Science in Electrical Engineering and went on to do a PhD. He then went on to become a professor and taught at MIT for over 45 years.

The catalyst for him to pursue his interest in speaker technology and sound engineering came from a major disappointment. To reward himself for earning his degree, he decided he would buy a state-of-the-art sound system. True to his engineering background, he worked through all the specs of what was available and made his choice – which he was bitterly disappointed with.

Inferior sound quality for his beloved classical music just wouldn't do! He set to work researching and building something better. He wanted to reproduce the sound experience that people had in concert halls, but delivered instead into their homes. This led him to his first breakthrough: that most of the audio experience in a concert hall was **indirect sound** reflected off walls.

Glenn Rifkin describes in his *New York Times* article how several idiosyncrasies and experiences that Bose and his corporation possessed uniquely enabled him to explore risky research pathways:

» He never offered shares to the public, so he had no one to please other than himself.

» He was following his passion, rather than money.

» His first product launch, the Bose 2201, was a disaster and his second product, the Bose 901, an unmitigated success. This gave him the feedback that he might not always be successful, but that it was worth persevering.

» He believed that better was often different, although this scared people, so he forged ahead even when people advised him to stop investing in an idea.

To market his 901 speaker (after learning from the failure of the 2201), he set about demonstrating it to the most influential journalists in the field of high-end audio. Once he had achieved several glowing reviews of his product, it began to take off. The audiophiles of the world led the way.

He also engineered a small computer to do a short sound quality demonstration to audio dealers around the country, convincing them to pay $1000 to set up a demo unit in their stores. Author Robert Ferris from CNBC goes into further detail on how this was quite an achievement in the 1960s.

Later, in 1972, witnessing that professional musicians were using his 901 speakers, Bose established a professional product division, which launched the Bose 800 and maximised sound quality with equalisation.

His next big breakthrough was inventing noise-cancelling technology, which found applications everywhere. The pilots of Voyager, the world's first non-stop, non-refuelled flight around the planet, were wearing his headphones. Commercial pilots routinely wear them these days, as do their passengers.

And on and on the inventions went. Acoustic wave technology ushered in a new era in sound quality; cars were equipped with Bose technology; and Bose developed more compact and even better speakers. Each time, the technology got more and more advanced. Bose found new ways to amplify live music, with his PAs and sound systems used at the Olympics. He even led an expert collaborative team that validated the debunking of original cold fusion experiments.

2004 saw a new business emerge, based on linear electromagnetic motors installed in car suspension systems to provide superior vibration control and performance. Tom Clynes, author of *The Curious Genius of Amar Bose*, describes the potential here for a passenger car to be able to corner like a race-car; for truck drivers to be insulated from debilitating vibrations; and for all drivers to deal with the most uneven road surfaces

Amar Bose passed away in 2013 at age 83. He gave the majority of Bose Corporation shares to MIT in 2011 as non-voting shares, never to be sold. His legacy is a corporation that always pioneered in everything it did, achieving outstanding commercial success on the back of original, ground-breaking technology and solving problems that he viewed as important.

Just like his headphones, he managed to cancel out the dissenters and went about creating something unique.

As Amar G. Bose himself once said:

> If I tell you that 'better' inspires fear – that even in the corporate world, people are scared of something better, you'd say that's ridiculous; everybody wants something better. Well, something better is always different. It isn't possible to make something better that isn't different. Whatever it is, if it's exactly the

same, it isn't better. So it's the 'different' that scares people. When something's different, it's a heck of a gamble. And that's where 'courage' comes in.

Bose Corporation is one of the purest examples of a ***product innovation*** differentiated company the world has ever seen.

Part Z

ACTION

Now you have a growing sense of possibility about how to disrupt, innovate and Xcelerate in your market.

You'll soon be itching to get into action, and this part of the book will show you how to do that. It will provide you with answers to questions like:

- » when should I use the Xcelerate framework?
- » what is the process for using it?
- » how can I actually launch a new business?

I know you take your future, and your organisation's future, very seriously, so Part Z is very serious. It's easy to talk a big game, but now you have to walk one.

As a current or future business leader, your last resort should be to cut costs through employee redundancies. That is just a proxy for failure to deliver good results because your strategy is flawed. If you're being parachuted onto the bridge and told to steer the ship, don't just fiddle around with some useless dials! You need to grab the helm and navigate the best course.

In this next part you'll learn how to develop potential new business units for growth. You'll choose which ones to launch and a process to launch them. This takes us from ideas and options to sales and results. But it will require shifts in how you approach changes like this today.

We are in business to grow, to expand the size of our markets and to provide purposeful work for our tribe. Not to shrink.

Part Z will show you exactly how you must do that, what risks you will face and what action to take – and fast.

2

Diving in

12

Now you've worked your way through Part Y, you have an understanding of what it takes to be an Xcelerator. I hope it's inspired some early ideas on how you could change your business to disrupt and innovate in your market.

But ideas alone don't impact your bottom line, and you're surely still asking yourself a lot of questions about implementing this stuff:

- » What is the best approach to apply the Xcelerate framework to my business?

- » How do I choose which model from the framework to change or work on?

- » What challenges am I likely to encounter?

- » What culture and environment do I need to achieve change and new growth?

- » How do I launch a new Xcelerator into my business and the market?

» Where and how do I start to actually implement this stuff?

There is a five-stage process, shown in Figure 12.1, that will answer the questions above and help you begin your journey to redefine leadership in your market:

1. **Locate**

2. **Generate**

3. **Activate**

4. **Iterate**

5. **Accelerate.**

Let's look at what each stage involves.

Figure 12.1: The five-stage Xcelerate process

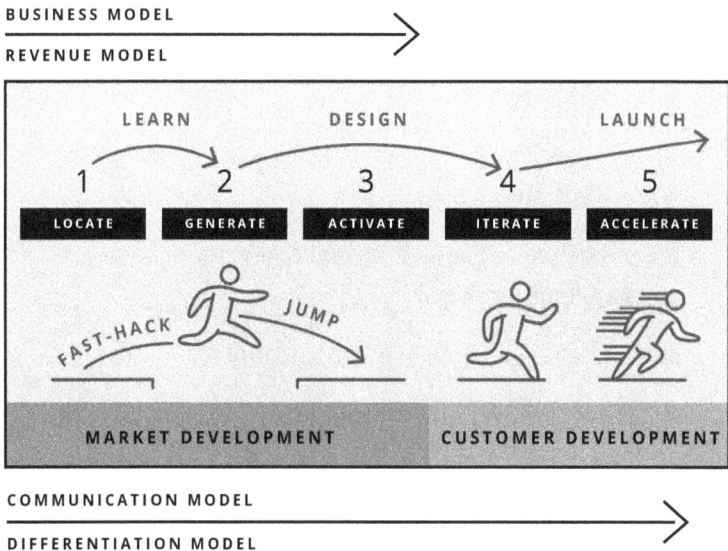

The Xcelerate journey

STAGE 1 – Locate

The first stage is to **Locate** where you're currently at and consider what's possible. A logical step is to get a team together and run a workshop using the Xcelerate framework.

Make this team as diverse as possible. Include people from outside your company and markets if you can, such as customers, suppliers, partners, mentors or facilitators.

This is about learning the framework and models as tools, with your team. Avoid having the 'That'll never work here' discussion. Keep your mind open. You're not testing the models, but rather establishing familiarity with them.

Look at other well-known organisations that have evolved over the past two decades, such as the examples provided at the end of each model in Part Y.

STAGE 2 – Generate

The next step is to **Generate** alternatives for your business and revenue models. These are the decisions made when starting a new business.

You will find that some logical options jump out when you ask what each of these would look like for your market.

Once you've done this, look at your communication and differentiation models. What changes could you make here?

STAGE 3 – Activate

Once you have generated one or two promising configurations that have the potential to shift markets and bring in new revenue, you need to switch from a market perspective to a customer perspective. That is, you need to design and **Activate** the details of your offering with specific customers in mind.

You need to determine:

- » Resources – Who is involved and to what extent? Who is on the team? What is their budget?

- » Support – Who are the sponsors? Where does the team go for help?

- » Details – What does the product/service look like? What are you selling?

- » Target – Who is the customer? What problem does the product/service fix?

This is where some of the challenges and dangers begin. (Sometimes they can start earlier in the **Generate** phase too). These challenges may include the following:

- » the rest of the business is probably operating a short-term time horizon – next month, next quarter – but this team will have a medium-term horizon

- » the team may get resources taken from them when the rest of the business gets busy

- » the bulk of the organisation may inadvertently stifle creativity, ideas or confidence with scepticism and

doubt – some of this change may threaten their world considerably

» the team will run into roadblocks that will require vision, courage and persistence to overcome.

You have, in effect, just created a startup.

STAGE 4 – Iterate

In the next stage, **Iterate**, you move away from designing to a testing phase.

You don't want to spend years developing an offering for your potential market only to find, once you've launched amidst great fanfare, that you missed what the customers were interested in.

This is a common failure in startups – discovering there was no market need.

In the **Iterate** stage, you take your concept to potential customers for feedback. They may want slight tweaks or wholesale changes, or you might uncover new insights from these conversations that take you in entirely new directions.

Remember, there are differences between market development versus customer development versus product development. Startups are not just smaller versions of larger organisations.

So here is where we can tap into the advantages of both worlds. We want to act as much like a startup as possible, but during this stage, access to your customers as an incumbent is a big advantage.

Two things are clear:

1. you must have an open mind to iterate

2. there must be enough substance and definition behind the offering for the prospects to be able to provide meaningful feedback.

STAGE 5 – Accelerate

Once you reach a decision about what you want to launch, you know it's still pretty raw and risky. What you need to do now is make another decision: to **Accelerate**.

Will you launch your new Xcelerator *within* the building or *outside* the building?

1. **within the building** – you will need a new team structure within the organisation for this to work

2. **outside the building** – you are launching a startup into a corporate accelerator, like a speedboat off the side of a battleship.

Both of these options require redeployment of people, time and energy.

Both options represent the start of your journey, of riding the wave of change.

───────────

This is where you must be bold, or else the effort you have put into talking about your change opportunity is likely to fail.

The only way forward is to launch – which is exactly what we'll explore in 13: Be a speedboat, not a battleship.

But first, let's put the five-stage process into practice.

I'm going to show you how easy it is to innovate your business model in under three minutes using the five-stage process.

How to choose?

*We complete the **Locate** and **Generate** stages on ALL FOUR models and generate ONE combination that becomes our Xcelerate framework for ONE new business unit.*

If there is a second configuration you are keen to explore, that will be a SECOND complete configuration for a SECOND new business unit.

*That is, once we enter the **Activate** stage we are working with at least one fully defined combination of choices in each model. For example:*

» *business model = **DIGITAL: creation***

» *revenue model = **UTILISATION***

» *communication model = primarily **Channels 2 and 4***

» *differentiation model = stays as **Customer Focus.***

How to innovate a business model in under three minutes

Yep, it really is possible to work out what type of business model you have in under three minutes.

The best way to prove this is to show you how.

Let's walk through an example of business model innovation together, using the following case study.

THE CASE STUDY

Professional services businesses such as medical practices, legal firms and accounting businesses are about to be disrupted.

It might be the early days, but there is evidence everywhere that advances in technology are taking business away from people and companies that don't embrace the changes.

If you are in a service business, then it's time to think about how technology will change the way your industry works.

As we saw in 3: King tides of change, access to knowledge is easier, it's democratising, it can leap country borders easily, **and** it's being automated.

The internet and cloud came first for *product-based* businesses selling through **PHYSICAL** assets; now it's coming for *services* being sold from **PHYSICAL** assets.

Consider the following:

» GP2U is an online medical practice in Australia

» BakerHostetler, a legal firm in the US, has hired ROSS: an AI lawyer that speaks in plain English,

can understand questions posed verbally and will instantly sort through thousands of previous cases.

» Smacc is using artificial intelligence to automate accounting for its clients. It's a startup, and many more will follow this early entrant.

It surprises me that many businesses I work with still refuse to consider the alternative business models available to them: models that have been enabled by new technologies or demanded by changes to buyer behaviour.

Architects, management consultants, engineers, IT consultants and financial planners are all in the same situation – there is both a threat and an opportunity to Xcelerate.

If you're in the service industry, what can you do before your future sales are impacted?

THE CHALLENGE

Let's see what business model options are available to professional service businesses.

To do this, we're going to apply the five-stage Xcelerate process:

1. **Locate**

2. **Generate**

3. **Activate**

4. **Iterate**

5. **Accelerate.**

Stage 1 - Locate

To work out what kind of current business model you have, return to the page you earmarked in 4: Business model.

The collective business model for professional services businesses is exchanging their expertise (**KNOWLEDGE**) over a period of time or as a 'task/job' (***contracting***)

Therefore, their current business model is **KNOWLEDGE: *contracting***.

(See how easy and quick that was? In no time at all, you'll be doing this stuff in under three minutes.)

Stage 2 – Generate

So what are your options for a new business model?

Option 1: **DIGITAL: *creation***

Introduce a new revenue stream that packages up some of your specialist knowledge into an online product. No doubt there is something you do for your clients that is non-standard; perhaps there is some way you can help your clients more (something proactive, preventative and helpful). This could be a series of recorded webinars, some video training, or an in-depth program to follow to help your clients' life, health or business.

For a doctor this might look like a monthly insight into living better, disease prevention, warning signs, the intersection between western and eastern medicine or reducing stress.

For an accountant it might be industry benchmarking reports, a management reporting pack each month, a series on improving profitability via pricing, the top five ways you see clients growing their businesses or templates for measurement tracking.

Option 2: **SYNDICATE:** *connection*

What if you created a membership group (either online or in person)? Your clients could get access to special guests, seminars, priority services, ability to interact with each other (which clients usually love), additional materials or attention on a particular topic. Or you might lead a movement on a challenging issue where they are involved in workshopping solutions.

This business model could also create connections between different groups of clients: e.g. helping startups find funding from investors or organisations interested in their product.

In this business model you are setting yourself up to facilitate value generation between you and the members, but also among the members themselves.

Now, how do these two options look in our business model table?

Figure 12.2: Business model table

ASSET TYPE	activity			
	distribution	connection	creation	contracting
PHYSICAL				
FINANCIAL				
DIGITAL			1	
KNOWLEDGE				Professional Services
MARKETPLACE				
SYNDICATE		2		

Stage 3 – Activate

Pick an option to work with.

It's time to allocate some resources to this, which is where it gets a little bit risky.

Now is the time to check your mindset, because you're probably thinking:

- » 'This'll never work'
- » 'No-one else is doing it'
- » 'This change seems dangerous'.

In the **Activate** stage, that sort of thinking is the only danger.

If you put a team on this, chances are the rest of your organisation will keep thinking the same thoughts you are and act accordingly, meaning it'll all come crumbling down around you. This could stall, delay or undermine your change project.

In 13: Be a speedboat, not a battleship, we'll delve deeper into some of the risks and how best to structure your team to create the right environment for success. For now, let's imagine the team will need space to operate (separation from the rest of the organisation), resources to get traction and sponsorship and support from the top.

In this stage, you are designing your new offering to the point you are ready and it is ready to be introduced to customer conversations.

Essentially, you've now created a new startup business unit.

Stage 4 – Iterate

Now it's time to take our new business and speak to a customer about it. Quite a few of them, in fact.

We don't want to do a large-scale launch until we have done that.

We have a startup business unit, and so we want to follow a lean startup methodology of getting feedback from clients.

Your customers may love or hate your new idea, but it doesn't matter because this is about getting valuable feedback. We are still designing our business and adapting our offering.

What you introduce to your customers needs to be clear, distinct and detailed so they can formulate a response. If you go to them with something that is half-formed, unclear or poorly developed, then their feedback will be less specific and less helpful.

Stage 5 – Accelerate

Imagine your customers like what they hear and want in – great! Now it's time to build your new business unit into your organisation or launch it out into the market in a different way as a stand-alone startup business (again, we'll look at this in more detail in 13: Be a speedboat, not a battleship.)

The offering may need to be built, fleshed out, expanded or scaled at this stage.

This is when additional resources in the form of money, external skill, time or a broader section of the business may need to get involved.

The idea is to accelerate this unit to market.

*Want to see how a company
innovates all four models:
business, revenue,
communication
and differentiation?
Download a worked
example of how
Netflix has evolved
from my website:*
paulbroadfoot.com/resources

———————

Use the right tool at the right time

You may still have some questions about how to do some of these things, like fleshing out your offering to customers in **Activate** or having the conversations with customers in **Iterate**.

So let's look at some world-class tools that complement the breakthroughs in ***market innovation*** that the Xcelerate framework delivers.

There are some very cool methodologies and tools to help companies that are facing the challenge of launching a new business.

The Xcelerate process and framework in this book is just one of them, but it is complementary to the other tools available, some of which you may already be very familiar with.

Three of the most well-known and powerful tools are:

1. Lean Startup

2. Business Model Canvas

3. Design Thinking.

How are these similar or different to each other? And how do they compare to the approach we have just walked through in this book?

The easiest way to explain this is to lay it all out in a table for you, as in Figure 12.3.

Figure 12.3: Differences between innovation tools

	Design Thinking	Lean Startup	Business Model Canvas	XCELERATE FRAMEWORK
Goal	Solving known problems	Successful startups	Business model generation	New growth for incumbents
Innovation type	Problem	Product	Strategy	Strategy
Innovates	User experience	Startup business	Business model	Business model, revenue model, communication model, differentiation model
Innovation outcome	Product	Startup with a successful product	Visual business model	New growth
Complexity and proficiency to use	Medium	Low	High	Low
Business stage	Any	Start	Any	Start and maturity
Development focus	Product and company	Customer	Customer	Market
Iterative	Yes	Yes	Yes	No
Speed of generating options	Fast	Medium	Medium	Super fast
Initial process (divergent/ convergent)	Divergent	Divergent	Divergent	Convergent
Famous for	Solving wicked problems, elegant solutions	Minimum Viable Product (MVP)	The Canvas	Finite frameworks

So how do you know which tool to use at which time in this five-stage process? Figure 12.4 shows you which tool to use when.

Figure 12.4: Which tool when?

Stage	Tool	Why?
Locate	Xcelerate framework	
Generate	Xcelerate framework	Market innovation
Activate	Xcelerate framework, Business Model Canvas and Design Thinking	Begins linking 'product' to customers and helps solve known problems in doing so
Iterate	Lean Startup Methodology	Customer innovation
Accelerate	*speedboat* (more on that soon)	

These tools work really well as a *portfolio*.

Each has great depth. Each has elements that overlap the boundaries I have placed around them, but we need a place to start.

Concentrating on just one is not the way to be successful at this.

How to pick your priorities

It's likely that in **Stage 1 – Locate**, and **Stage 2 – Generate** you will identify several different alternatives. So how do you pick and prioritise which to work on first?

1. IDENTIFY THE TRENDS

    a. take a close look at trends beyond your *marketplace* (e.g. those described in 3: King tides of change)

    b. take a close look at trends in your *industry* (e.g. mobility)

    c. take a close look at all emerging *technology* trends (e.g. sensors)

Once you have identified these trends, they should highlight threats and opportunities to your current business model and your new alternatives.

For example, king tide 1 describes access to knowledge being democratised. Feeding into this is automating information processing by artificial intelligence. These together will attack any business models with **KNOWLEDGE** as the income-generating asset.

2. ASSESS YOUR COMPETITION

And by this, I don't mean worry about your competition; I want you to map all of your competition into each of the four model types in the Xcelerate framework.

Identify where they all sit. From this you will see white spaces and that's where you want to go.

You should define 'competition' not simply as companies just

like you – it can be any competing alternative for your offering. For example, a telco might define 'competition' as other telcos, but in reality, Facebook Messenger is a service that directly competes against a telco's text messaging services.

3. LOOK WITHIN YOUR MARKET AND OTHERS

Look for:

a. Big players that seem to have very similar offerings (Australia's Big Four banks, Australia's two major food retailers)

b. Low customer engagement ratings or businesses known for unhappy customers

c. Industries that are regulated and have slow pace of change (taxis, Uber)

d. Changing consumer/customer behaviours at the fringes.

Decisions, decisions

How easy is it to change each of the four models: business, revenue, communication and differentiation?

How much work is involved, and when?

Take a look at the axes in Figure 12.5.

Figure 12.5: The Xcelerate Framework

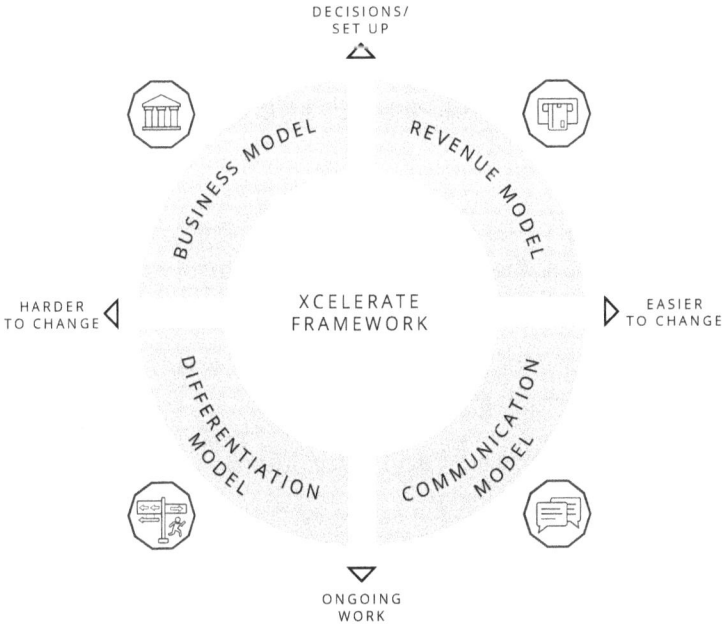

DECISIONS/
SET UP

BUSINESS MODEL

REVENUE MODEL

HARDER
TO CHANGE

XCELERATE
FRAMEWORK

EASIER
TO CHANGE

DIFFERENTIATION MODEL

COMMUNICATION MODEL

ONGOING
WORK

The vertical axis shows us that the business and revenue models tend to be decisions made at the formation of a startup, business, company, organisation or enterprise.

Makes sense, right? When we set up a company, we have to decide our **ASSET** and *activity* at the outset and determine how we are going to charge and receive revenue from sales. We need this from the very first sale. Thus, our business model and revenue model are established early.

Conversely, how a company communicates (communication model) and how it differentiates (differentiation model) tend to involve significant and ongoing work.

If we choose to use direct sales in our communication model, we need to continuously invest in training, hiring and managing this.

If we choose Customer Focus as our differentiation model, we will need to constantly strive to get better at servicing our customers, listening to them, understanding what they want and training our teams to service them. Lots of ongoing work right there.

The horizontal axis indicates how easy or difficult it can be to change the models. Business and differentiation models are like the DNA of companies once they grow large. There are often lots of sunk costs in the business model and lots of resources to shift if we want to change it.

The differentiation model is a bit different. If a company has a strong differentiation model, they are focusing on one of the four types (market, product, company or customer). It is less likely for this to be changed. What is much more likely is that a company has not clearly identified which type of company they are; they are likely trying to be more than one. In this case, the change required is deciding which one is right for them, and that will require more ongoing work.

Conversely, it's easier to change our mix of communication model channels or change our revenue model from TRANSACTION to UTILISATION, for example.

Next, you will learn how to use the five-stage process – regardless of which model you have just innovated – to launch a new business unit.

Xcelerate now

1. There is a five-stage Xcelerate process to create a new business unit:

   i. **Locate**

   ii. **Generate**

   iii. **Activate**

   iv. **Iterate**

   v. **Accelerate**

2. The Xcelerate framework should be used in the **Locate**, **Generate** and **Activate** stages. Other tools, like Design Thinking and Lean Startup, are used in the **Activate** and **Iterate** stages.

3. To help pick your priorities, use trends like the king tides and look to your market for areas where there is no current offering, low customer engagement or where company products and services are very similar.

4. The revenue model and communication model in the Xcelerate framework are easier to change than the other two models.

5. The communication and differentiation models require a lot of ongoing 'work'.

5

Be a speedboat, not a battleship

13

Ken Warby holds the water speed world record with a speed of 511.11 km/h. He started building *The Spirit of Australia* in 1972 in his good ole Aussie backyard using three power tools, a drill, a belt-sander and his hands.

He first set the record on 20 November 1977, at 464.46 km/h, and then broke his own record again on 8 October 1978. He has held this record for almost 40 years.

Approximately 85% of people who have attempted the water speed record since 1940 have died trying, making it one of, if not the, most dangerous sporting quests in the world.

What makes this even more remarkable is that Ken's super-fast boat was built with very little money.

In fact, in 1974 he ran out of money for his project. He threw in his day job as a Makita salesman and dedicated himself to his boat. He began travelling around service stations and supermarkets to raise money, displaying his unfinished boat and selling oil paintings.

Two of his first sponsors were Shell and Fosseys (you may remember this retailer, which later became part of the Coles Myer Group). Bringing the additional sponsors on board gave Ken the money to finish the boat.

Startups, like Ken's, are famous for having vision and passion in spades, an audacious goal and the guts to go 'all-in'. But there were a couple of other crucial elements to Ken's success:

1. He teamed up with a couple of buddies from the RAAF, Crandall and Cox, who helped renovate the Westinghouse J-34 turbojet engine (which Ken had bought for under $100 at a RAAF surplus sale) to working condition.

2. He met Professor Tom Fink, dean of the faculty of engineering at the University of Sydney, and built a scale model that he tested in the university's wind tunnel, the results of which impressed Tom so much that he joined Ken's crusade.

These two injections of expertise greatly increased Ken's chances of success. Ken completed his first design in 1970 on his kitchen table. It took him seven more years to break the water speed record.

He reached the **Activate** stage (from the previous chapter) in 1972 and his **Iterate** phase went for five years.

During the **Iterate** phase he tested the boat on water, increased his speeds and continually tweaked his design.

The key questions this raises for 'the startup' are:

» Could Ken have achieved his goal quicker if he was well funded (like a corporate)?

» Would he have been successful without his RAAF buddies, the wind tunnel for testing and the professor (expertise and ecosystem)?

Startup versus enterprise

Let's reverse that thinking.

What would happen if we tried to 'manufacture' this *speedboat* from within corporate ranks?

» How many times would the idea we're launching have been challenged to be shut down because it was taking too long?

» How would the external expertise and ecosystem have been introduced successfully?

» How could you possibly get a team to care as much as Ken did about this 'new business' to sustain it through setbacks and challenges?

Figure 13.1 shows us the *speedboat* characteristics of both startups and enterprises. The middle column presents the BEST of each. It shows the ideal type of *speedboat* we want to build if we are going to set world records.

Figure 13.1: Startups versus corporate speedboat

	Startup	Hybrid – best of both worlds: a *speedboat*	Enterprise
Passion and engagement	Ultra-high (can be evangelical and delusional)	High	Medium to low (can be disengaged and disillusioned)
Risk appetite	High	Assisted/ portfolio securitised	Low
Openness to change	High	Maintain high	Low
Speed of communication from customers to CEOs	High	Maintain high	Low
Distribution capability (to customers)	None	High	High
Business acumen	Low	High	High
Funding	External and often limited	Corporate sponsors of startup units	Internal and often adequate
Structure for innovation	Alone, incubated or accelerated	Accelerated	Internal or facilitated
Ability to scale	Low to medium	High	Medium to high
Network	Varies	Blended, diverse, extended	Low, but sometimes international

Speedboat specs

*There are three key characteristics of a **speedboat:***

1. *a startup's agility and passion*

2. *a corporate's resources and funding*

3. *an ecosystem with both the skills and diversity to spark breakthroughs.*

How to launch a speedboat

Great! So how do we build and launch a hybrid **speedboat**, you ask?

In 12: Diving in, we highlighted two ways to go about this: *within* the building and *outside* the building.

1. **within the building** – you will need a new team structure within the organisation for this to work

2. **outside the building** – you are launching a startup off the side of the battleship into a corporate accelerator.

This conversation is going to feel uncomfortable for career executives in enterprises that have never been exposed to the startup space, the VC landscape or current best practices such as the tools we described in 12: Diving in. But, hopefully, now you've read Part X you know WHY there is a need for you to tear down your castle walls, disrupt your mindset and have a crack at fast change.

Let's not sugar-coat this. It is incredibly hard to achieve change in any organisation. The bigger the organisation, the harder it is.

Humans are hardwired against change as it is, but in a large organisation you've got intense sign-off processes, tacit and direct approvals required to alter things and employees who may not always be engaged or on board because of the diffusion of communication through managerial layers. Departure from the way things have always been done just represents risk, so yes: it is tough.

So when we say there are two ways to get a ***speedboat*** going, we are expecting it to be a big departure from your comfort zone.

Both approaches have risks, and both have unforeseen icebergs you will need to navigate. If you launch internally, you will need to create a team structure you haven't had before. It will require definition and tremendous conviction to change.

If you launch it externally in a corporate accelerator, it's even less familiar, but you mitigate the three main risks of launching internally that we're about to explore .

Sure, there are other risks or icebergs if you go external. Australia's incubator and accelerator space is expanding beyond its experience, being propagated by government money that is less than discerning, and is creating a bubble of its own. But the best accelerators will outperform an internal approach. Here's why.

Make or break

When the First Fleet arrived in Australia and landed in Botany Bay, they found minimal food and shelter. The convicts were not trained in farming and the soil was not conducive to growing the crops they had brought with them. They didn't enjoy local sources of food, were poorly skilled fishermen and couldn't communicate with the locals. Their inadequate tools made building and construction using hardwood trees near impossible.

For two years they struggled as a colony, enduring hot dry weather and drought. They despatched the ships 'Sirius' and 'Supply' to bring food back, while they tried to survive on rations. Some tried to escape the colony and make it on their own, but they either perished or returned unsuccessful.

These difficulties were unexpected. Had the Fleet been more open they could have learnt a great deal from Australia's first people, the indigenous inhabitants, who were perfectly comfortable living off the land.

When we try to do something different in business, we often endure early difficulties. Sometimes these hardships force us to turn back, or we adapt and

overcome. The early settlers did it tough and may
not have made it without the extra food that arrived
by ship.

The landscape that confronted the early settlers was
always going to make or break them. In the same
way, your current business landscape will provide
challenges that if unsolved or unresolved can create
struggle. The context in which you work to launch
something new can make or break the outcome.

The three internal icebergs you must navigate

There are three major risks, or icebergs, that threaten the launch of **speedboats** internally:

1. culture
2. structure
3. ecosystem.

As we explore each one in detail, I want you to think about this: the *Harvard Business Review* article 'Organisations can't change if leaders can't change them' cites studies that 60–70% of change initiatives fail in business transformation.

It follows that if you want your changes to the Xcelerate framework to have a better than 30–40% chance of success, you need to do something differently.

You've seen how Amazon, Netflix, Apple, IBM and many other companies have done it in this book already. So we know it's possible.

Additionally, research tells us that global employee engagement averages about 65%. In Australia, we are in the 55–60% bracket (Aon Hewitt's *2016 Trends in Global Employee Engagement*). That's 40–45% who are not engaged. (There are two ways to look at this: (1) it's going to be hard to achieve new change if people are not on board when you launch a **speedboat** internally; or conversely (2) the **speedboat** may be just the initiative needed to get some of those wanting change back on board.)

Navigating these three icebergs can be done! But you must get thinking about how to do it, which is what we'll explore now.

Internal iceberg 1 – culture

Reed Hastings ran another startup prior to founding Netflix. He sold his stake in this original startup for about US$750 million. Why?

He said he was elevated beyond his management capacity for the size of the organisation it had become. (He is probably selling himself quite short, but he was obsessed about getting it right at Netflix.)

His 124-slide deck on Netflix culture (Google 'culture Reed Hastings') is the product of years of evolution in his business. Facebook COO and author of *Lean In* Sheryl Sandberg said it 'may well be the most important document ever to come out of the Valley'.

As Reed Hastings describes in the slide deck, as companies grow they reach a point where they need some systems and processes to reduce the chaos. And that tends to be when the bureaucracy of running an ordered business starts to stifle being nimble, creative and decisive.

The culture Reed Hastings describes may not suit all companies, but it's THE standout example of a great approach to bringing their culture to life by answering the tough questions such as 'What to do with brilliant jerks?' It's clear who fits and who doesn't. It's real and in many places unexpected, such as unlimited vacations and how to stop the best employees leaving when the business gets larger.

One reason why business innovation can fail in larger organisations is that the decisions and activities required are incongruent with the current culture.

———————

It may be due to demanding risk-taking in a culture of conservatism, or employees could be fearful of experimentation, discouraged by reactions to failed launch attempts in the past.

Reed Hastings has 'baked in' risk-taking, proactivity and entrepreneurial endeavour into the Netflix culture. If, as is more likely, your culture is not like this, it may be best to have more 'separation' of your **speedboat** from day-to-day operations. Cultures are essentially the personality of the organisation, and personalities can be hard to change.

Separating, protecting and supporting the 'new bits' is really important, as the existing culture can act like white blood cells attacking the new 'infection' to remove it from the organisation.

Another, more radical, way to adapt culture to foster innovation is described in Frederic Laloux's book *Reinventing Organizations*. He describes the evolution of organisations through the ages, while at the same time colour-coding them.

Under his system, a 'Teal' organisation has no formal, fixed hierarchy. A Teal organisation is self-organised, self-managed and decentralised. It is hard for most people to imagine a company with no leadership and management positions.

The Morning Star Company, a large tomato processor in the US, is an example of a Teal organisation. There are no bosses. Each person develops a 'Colleague Letter of Understanding (CLOU)' with the colleagues most impacted by his/her work, and they aren't 'empowered' because there are no senior people to empower them. The Morning Star Company, owned by Chris Rufer, handles about 30% of California's tomato crop and at last count (being a private company, data is limited) they were to their way to $1 billion in sales revenue.

In a *Harvard Business Review* interview, Gary Hamel suggested to Chris Rufer that he had cracked the problem of how to run a company without managers. Rufer countered with:

> Everyone's a manager here ... We are manager-rich. The job of managing includes planning, organizing, directing, staffing, and controlling, and everyone at Morning Star is expected to do all these things. Everyone is a manager of their own mission. They are managers of the agreements they make with colleagues, they are managers of the resources they need to get the job done, and they are managers who hold their colleagues accountable.

Again, there are no bosses, no positions, no position descriptions.

And there is order, profitability, innovation, flexibility and happy campers.

Whether you are a Teal organisation or not, you need to be flatter, decentralised, more collaborative and more democratised.

It seems strange when you sit and reflect on it. As *information* flows up the hierarchy, it gets filtered and reformatted, and as *decisions* flow down the hierarchy we assume that those at the top have better input and make better decisions than those lower down. But it's not necessarily so, I would argue.

The returns of diversity, whenever they are studied, are clear. If we discount certain voices and overemphasise others, we are not going to see shifts, innovation and new business creation in our enterprises.

Navigate the icebergs

» *Give your employees more ownership. Let them decide, allocate, and finish projects.*

» *Form diverse teams of no rank.*

» *Encourage a culture that rewards performance, not position.*

» *Ensure there are shortcuts to the hierarchy for information to flow faster and be less filtered.*

» *Make sure every piece of 'control' and 'system' is necessary, or else remove it completely.*

» *Embrace experiments and freedom to be truly innovative and creative.*

Internal iceberg 2 – structure

The structure outlined below acknowledges there are different activities to be managed when working to innovate your market via new businesses. There are the day-to-day activities, the activities of the 'new bit' and the activities that take the new bit and scale and integrate it into your whole organisation.

In a blog article called 'What every Institutional Innovation Program gets Wrong', Bud Caddell of the consultancy NOBL describes a structure of three **internal teams** similar to the one below. I have expanded the theme here.

1. **BATTLESHIP**
   This is the go-to team for current operations. They must keep the home fire burning – it cannot go out. They are in charge of current business as usual, and about 60% of budget. Their focus is the short-term horizon.

   CHARACTERISTICS OF THE TEAM MEMBERS:
   » they are happy where they are

   » they do not like change much

   » they are more attached to existing ways of doing things

   » they are experienced and skilled.

2. **SPEEDBOAT**
   This is a team of experts in methodology tools or with access to expertise such as the Xcelerate framework. They need freedom to explore outside the building. This team should hand off 'startup projects' with some early-stage feasibility to the

cruiser team. They handle 20% of budget and focus on the medium and long-term horizon.

CHARACTERISTICS OF THE TEAM MEMBERS:

» they are more entrepreneurial; at risk of leaving if not challenged

» they are constructively rebellious

» they are passionate, creative ideas people

» they are critical of opportunities that are not explored

» they are curious and fascinated by what is possible

» they love change

» they are prepared to challenge and be open minded.

3. CRUISER

This team is responsible for integrating successful **speedboats** back into the battleship. They handle 20% of budget and focus on the medium-term horizon.

CHARACTERISTICS OF THE TEAM MEMBERS:

» they are 'people people'

» they have high levels of respect and influence

» they need to be convinced to get on board, but are determined and tenacious once there

» they measure change practically and work to break down resistance constructively

» they improve the overall solution as part of the integration and implementation process.

Having these three teams in place ensures you have a *portfolio* approach to your business. This ensures always having *businesses* at different stages in your growth pipeline, just as you would always have products in your innovation pipeline.

Internal iceberg 3 – ecosystem

No matter what you read on implementing innovation, you will find a handful of common elements. These same elements are recommended for leadership of others, change management and pretty much every other business initiative.

They are:

1. get out of the office (to get exposure and insights directly)

2. give people freedom and ownership so they are engaged

3. collaborate and be diverse, as it yields better outcomes

4. support and guide, but let the team create

5. learn and never stop there.

The *New York Times* (NYT), as with all newspapers around the world, including our own, have worked to reinvent themselves in the face of falling paper sales and falling advertising revenue.

During one of their phases of reinvention, the NYT pushed to become more entrepreneurial. They hired John Geraci, a successful entrepreneur, as Director of New Digital Products. His hire and the hire of many others into the newspaper was about introducing more entrepreneurial DNA into the

business. This evolved into a desire to adopt a VC mindset and a special executive board was created to evaluate opportunities, allocate resources, and evaluate risk.

Many companies in Australia are doing the same thing right now.

Lean Startup, MVP and innovation hackathons are all the rage.

In an article in *Harvard Business Review*, Geraci positions his two years at NYT as a learning era for both himself and the organisation, with two of their three new products relegated to free offerings and one shut down entirely.

Their goal to create new revenue sources has failed for the moment. Geraci attributes this failure to lack of entrepreneurial *ecosystem*.

Silicon Valley is the world's best ecosystem for innovation. Here is a list of some of the companies to come out of the Valley over the years:

> Adobe Systems, Advanced Micro Devices (AMD), Agilent Technologies, Alphabet Inc, Apple, Cisco Systems, eBay, Facebook, Google, Hewlett Packard, Intel, Intuit, Juniper Networks, Lockheed Martin, Netflix, Nvidia, Oracle Corporation, Salesforce.com, SanDisk, Symantec, Tesla Motors, Yahoo!

What is it about Silicon Valley that has 'worked'?

My answer is that it's the network effect of having investors, academia, startups, tech skills, repeat entrepreneurs, mindset, and VCs all in the same locale, all looking to learn and grow.

Startups understand **ecosystems**.

They 'get' that they need help, they know they need money and they intuitively understand that interaction with others outside their world leads to breakthroughs and learning.

Larger organisations get this too – but how do you pull it off when you are bigger?

Just like the NYT, Australia's market incumbents are trying to crack the code of business innovation. In some cases, companies are building their own hubs and coworking spaces, while in others they are bringing in startup specialists or actual startups to 'add entrepreneurial DNA', as Geraci puts it. Some of the tools mentioned already, such as Design Thinking, are getting a turbo thrust.

Australia's banks get it; they can see the impact that FinTech will have on their business model. They are first movers on innovation experiments.

That's why you need an external boat ramp – an accelerator

The ecosystem iceberg and risk we've just explored is very hard to mitigate when you launch internally. It's really hard to get external ecosystem inputs into organisations, unfiltered and unfettered.

We know from our discussions about culture and organisation structures that in many cases your current organisation may not be fertile ground for the kind of breakthrough innovation required in a world that is outpacing traditional, hierarchical, incumbent rates of change.

One big answer to corporate innovation is not too dissimilar to how startups launch. The concept is to launch a **speedboat** from the battleship, to chart new growth opportunities, out of the day-to-day reach of that corporate bureaucracy – an immersive, diverse and supported **speedboat**.

Launching a **speedboat** team into the right corporate accelerator ensures the culture, the structure and the ecosystem risks are largely removed.

There are numerous ways to engineer this **speedboat:** everything from 100% equity to a sponsored low-equity startup. The investment vehicle ownership structure is a discussion for another day. Suffice to say one straightforward option is to have a team of employees in an accelerator. It's easy and it's a bloody good idea.

An accelerator offers much more structured support as well as timelines and milestones to meet, usually. Significant mentoring occurs and time periods are generally short (three to six months), although this varies.

Why a corporate speedboat is like having a gap year

This decision resembles the one faced by some young adults : whether to take a gap year after finishing school. We know that if they travel overseas, they'll be exposed to different cultures, experiences, and greater risk than if they stay in Australia and study. And we also know it'll be a growth experience for them.

Sending employees out in a corporate **speedboat** to work with others in an accelerator is just such a decision: greater risk, greater growth.

So you've nearly made it through the book. We've been on a long journey together. We've seen why walls don't work, explored the Australian research into our ASX companies, seen the king tides of change, and of course, looked at how you can become an Xcelerator and how you need to launch a **speedboat** off the side of your battleship if any of this is going to work.

Has this been done before?

Yes.

Did it work?

Yes.

What did it look like?

The following section, 14: Fast forward with Melanie Farmer, is an interview I conducted with the leader of a highly successful program run by Sussex Innovation Centre in the UK over ten years ago (this shows you how far behind we are in Australia!). This program first coined the term 'lifeboats' for their corporate innovation programs. (A fitting name and one I iterated on in this book as a euphemism for what they may actually be.)

This last section of the book will show you – and hopefully inspire you about – what is actually possible. Because it is all possible. Nothing you have read is the stuff of fiction: it's all real. All you have to do is grab your lifejacket and jump on board.

Xcelerate now

1. Both startups and enterprises have drawbacks when it comes to launching. We want to build a hybrid ***speedboat*** to enjoy the best of both worlds.

2. There are two ways to launch a ***speedboat***

    i.   internally

    ii. externally.

3. There is risk and discomfort in launching a ***speedboat*** – but get over it, grab your lifejacket and jump in the driver's seat!

4. If you launch a ***speedboat*** internally, you need to navigate three risks, or icebergs:

    i.   culture

    ii. structure

    iii. ecosystem.

5. If you launch an external ***speedboat***, you do this in a corporate accelerator.

*Fast forward
with
Melanie Farmer*

14

Melanie Farmer is the Associate Director for Research Engagement, Development & Innovation at Western Sydney University.

Melanie started her working life as a primary school teacher in Australia prior to moving into corporate sales and training. From here, she ran her own theatre company in Singapore, then moved to the UK where her digital startup grew, allowing her to exit and join the award-winning Sussex Innovation Centre. Founded in 1996, the Centre is one of the oldest business incubators in the UK. Here she led the innovation team, which supported companies inside the building, corporates outside and SMEs in the area.

Melanie managed innovation partnerships involving as many as 17 partners in 11 countries across the EU. Holding an MSc in Management of Marketing for Innovation, she drove innovation projects leading to outcomes such as doubling revenue for firms within 12 months, shaping up firms for successful sale of the company and streamlining many innovative firms to enable them to scale up.

In addition to helping others take their inventions to market, she herself developed and grew the innovation services of the Centre and was pivotal in securing several of their awards.

How did the innovation landscape compare between 'then' in the UK and 'now' in Australia?

Some of the discussion feels like I am going back in time ten years. There is a lot of finger-pointing. It's the job of the industry to go and find inventions and universities, but then there is, 'Well, yes but we think it's the job of universities to go out and find industry and talk and engage with industry'. So that reminds me of the conversations that were happening more than ten years ago in the UK.

I think that a lot of countries – Japan, USA, UK, parts of Europe and Australia – have all suffered from the idea that what we need is startups, and so we go and build an innovation centre, we fill it with startups and then we don't support them with the right team to fill the gaps in their knowledge. Within five years they're out of business, and so the government of the day gets to count the jobs and the companies that they've created, but those are not lasting.

That's been a failure of many governments – not looking long-term – and I think that corporate innovation is a really good way to make sure that the problems we're solving with these startups are things that the world needs. A big corporation is generally investing in big problems, such as banking or agriculture.

How do corporates and startups compare in their approach?

Let's compare a corporate and a startup to an ocean liner and a lifeboat. The corporate is an ocean liner trying to change direction or disrupt itself, which requires abandoning the safety checklist and risking the possibility of running into an iceberg.

Trying to innovate inside is a very difficult thing. Many corporates inadvertently stop themselves from innovating or encouraging an innovative culture through their own processes.

If you have an entrepreneur inside a corporate, often they're not happy, it's not the culture for them, they want to leave and there are many famous examples where they do leave.

What were some of the concerns of larger companies when sending key people off for six or 12 months? Were they saying 'How are we going to know they are on track? How are we going to know we are getting a return on our investment?'

They did have concerns about return on investment. Saying that, this was not the only metric for success. They were actually interested in understanding the innovation process and in building closer links with the Sussex Innovation Centre, which at the time of starting these programs was already an award-winning innovation centre.

They wanted to understand investor networks and how they work. They were taking a risk and said, 'We're putting this in the risk basket, we're trying it, we're experimenting' – which one has to do, in my view, because this is a high-risk activity with a high failure rate.

The culture and flavour of these lifeboats was really: 'Let's throw people into the lifeboat and see what happens'. Return on investment was a bonus and, although that clearly came, it wasn't the number one, front-of-mind focus for them.

The Sussex Innovation Centre states that typically only 15% of companies go on to be profitable, but the Innovation Centre's success rate shows that 85% become sustainable businesses and one in six achieve turnover in millions.

———————

How did funding work, and what were the challenges?

When we started out as an initiative, we had our own investors made up of alumni from the University and various other sources. This small group of investors grew and grew, but we faced the challenge of legal implications if we were seen to advise investors whether they should put their money in the companies we were building.

We were not the only ones facing that challenge. We had two other groups in south-east England who had those concerns with their own networks. The three groups came together, and we decided to combine and house our list of investors with the South East England Development Agency (SEEDA), a regional government agency. SEEDA then administered the resulting program: the South East Capital Alliance (SECA).

The SEEDA was impartial, so it would send a call out to investors from these three combined lists announcing, 'We're going to showcase environmental technology in March. So if anyone's interested in investing in green tech, come and see the pitches'. And so on for other months with each, as best they could, focusing on a particular sector or interest area for the investors.

Not only did that triple the size of our investor network, it worked better for the investors because they could pick and choose where they wanted to invest their time as well as their money.

What's your advice on how a corporate manages risk?

I think that in Australia, we're trying very hard to de-risk everything and actually we have to take risks. This idea that you can measure and monitor and pick winners is great, but I think this is completely the wrong approach. I think that we have to measure how much risk can we afford.

If we're going to go and gamble, if we're going to take a risk, then we should know how much we can lose and what we can bear in terms of a loss.

If I were running a business, I would expect that 50% of my money and my time and my energy would go into strategically building my operational proficiencies in my core business. Then I would break up 25% into what I know is coming, so I'm ready for it, and then I want 20–25% where I'm doing 'innovation' (or 'crazy things').

I'm creating the future, I'm looking for problems I didn't know were there and, in this last 25% of my resources, I'm more likely to make mistakes and lose money. Design thinking and the right support and analysis can mitigate much of the risk factor here, but it remains that innovation is – for the most part – risky.

So how do you build a hybrid between a risky startup and a risk-averse corporate?

I think a corporate can benefit from putting their people in an external innovation centre because they know that the team in there – if they've picked the right centre – will provide the right, timely support to help them avoid making basic mistakes. Conversely, expecting their staff to innovate inside a corporate (unsupported) or putting them in a corporate innovation lab may leave them vulnerable to making some of those basic mistakes.

In our lifeboat model, there would be an executive sponsor from the corporate who will drop in every now and then to see how things are going. The group with the lifeboat would join in with the entrepreneurial community in the innovation centre. We would have, at any given time, around 120 companies in the building and up to 500 tenants overall if you include our virtual tenants (working from home and not yet requiring office space).

What many Australian corporates are doing is setting up their own innovation labs, and I did see this happen in the UK. But if I were to go into those labs as an innovator, I wonder if I would miss the opportunity to be exposed to diversity?

By putting innovators into an innovation centre, they're surrounded by people from different sectors and they're surrounded by those who are breaking rules and being creative and taking risks. Innovation comes naturally when you get two diverse industry specialists together to solve a big problem: e.g. an economist and an anthropologist solving the housing crisis.

That doesn't really happen in my mind if you get the guy who works on the gas pipeline to sit with another guy on the gas pipeline and together they think about how they can make a better gas pipeline.

> *By keeping it all in-house,*
> *I think you miss a whole lot*
> *of happy accidents.*
>
> ---

The strength of taking your people out of your company and putting them somewhere else, where they are completely out of their comfort zone, is that they are new again, and can kick-start their brains more easily.

Through the lifeboat concept, you're also able to retain intrapreneurs (internal entrepreneurs), because before that they were leaving. Why would they stay when they didn't have any mechanism, any outlet, for their talent? This was a way for corporates to retain their innovative people – their future competition, if you like.

Where do you see the opportunities in Australia?

I think there is a huge opportunity to innovate in rural Australia; however, the people that are needed there, those 'teams around the entrepreneur' are not in rural Australia.

They're in our cities and on the coast.

For example, I'm in Sydney but I'll bet you someone in Alice Springs has a fantastic solution for drought. How are they going to get their idea to me, and how am I going to get my advice and support to them?

I don't think the internet is the answer.

I would love to see an innovation bus touring about, and I'd love to see face-to-face support happening in rural Australia, because I think that is where the real solutions are for big problems in the world. I have this feeling that in the cities, we are somewhat disconnected from agriculture, drought and fire and instead of solving those big issues, we invent cool gadgets or apps (caveat: granted, some of these are actually life-changing).

Can you see anything else missing in our approach to innovation in Australia?

I really think that in all of this, citizen science and citizen innovation really needs a kick upwards in Australia. We are still talking about the triple helix here. The triple helix of innovation supposes that innovation is driven by three players: academic, government and industry.

The quadruple helix includes a fourth player, the citizen. Citizens or civil society are the end user: the person who invents the new type of robot for Lego because they're a super-fan. The guys who built the mountain bike and adapted it because they personally wanted it to be able to go off-roading. These pro-amateur innovators are our taxi drivers, our nurses, our policemen. They're the people I think have solutions to our big problems, but they're not businesses,

they're not government, they're not academia so their ideas are left untended, unsupported.

I think Australia needs to actively include citizens in the innovation process, particularly in rural Australia.

*A lot of the disruption
that is occurring is not occurring
inside a corporate, it's not
occurring in the university, it's
not occurring in the lab and the
government isn't driving it.*

———————

It's happening from the guy who couldn't pay his rent who asked himself, 'What if I could rent my sofa out?' and grew that into today's Airbnb. It's happening from the citizen, and if we are ignoring the citizen then we're missing the opportunity to solve things in some unexpected and interesting ways.

CON
CLUS
ION

The company lifespan in Australia has halved in the past 40 years, and that is before the digital revolution has really impacted. What will the rate of change be like in the next five years? Seeking safety by redoubling efforts on optimisation and efficiency internally behind company walls does not work. As you have seen in this book, Borders, Blockbuster, Sharp, Tooth & Co and many others are all big names that lost their market fights and more.

Enterprises do not spend enough time innovating the way they work, nor do they focus on market innovation by seeking business, revenue and communication model alternatives. Further, they don't focus their differentiation narrowly enough, which makes them beige wallpaper to prospects and customers.

We need to innovate how we innovate. We need to take more risk to have less risk.

The fast, finite Xcelerate framework in this book gives us the breakthroughs for our strategy, but still more needs to be done to bring these to market. You need to set up your **speedboats** and launch them. You will be challenged by those around you, by those outside your organisation and by your own doubts.

When IBM first launched their services business in the 1990s, the press speculated it wasn't enough, and they might not survive. If you read IBM's quarterly results announcements 20 years later (after a stunning turnaround), you will hear the same commentary on their current strategy. Will it be enough? Will they evolve?

Xcelerator sales start small, but grow exponentially. Just remember that.

Call to action

I urge you to do three things in the two minutes, two days and two weeks after you read this book.

1. Two minutes

Decide what kind of leader you want to be. Are you a leader who focuses on doing everything required to manage today's business? Or do you want to be a leader who focuses on exploring the next level of performance, on the future, by taking some risk, on being first and being lonely? Decide now whether you will be:

☐ *an Xcelerate leader*

☐ *a traditional leader*

2. Two days

Rally your troops around and start thinking about your battle plan.

What trends might change the way:

- » *your customers buy?*
- » *your competitors compete?*
- » *end users experience your industry?*

3. Two weeks

Complete the **Locate** and **Generate** stages, ready for the **Activate** stage of the 5-step Xcelerate process.

Add up how much money you have budgeted this year for training and other activities in:

- » change
- » innovation
- » communication
- » strategic growth
- » sales
- » high-performance teams
- » leadership.

How much of that money do you suspect isn't delivering a return? How much of that money would you be willing to redeploy as a bet on one or two **speedboats**? A **speedboat** is like a crucible where the capability list above will be tested, learned and enhanced, every day.

MOVE IT

When you ride a motorbike and a truck suddenly pulls out in front of you from a driveway, your natural instinct is to turn the front wheel away from the threat – but that's going to get you killed. What you are taught to do instead is counter-steer. When the truck pulls out from the left, you turn the wheel *into* the threat. This drops the weight on the opposite side and enables you to follow an 'S' trajectory around the truck. This does NOT feel natural at first. It takes lots of practice until it does.

It's the same when it comes to business, and particularly innovating in business. Some of the best decisions you can make as a leader go against many of your natural instincts, like these:

1. When you are faced with changing markets and poorer performance, you naturally seek safety by sticking or returning to activities that have worked for you in the past. You seek familiarity and have a tremendous ability to rationalise any behaviour you choose to adopt. You hunker down behind your castle walls.

2. When you look at all your customers and potential customers and the primary ways you work to attract them to your business, you try to be all things to all of them. You don't sufficiently narrow your message, your activities, your focus. Narrowing is not a natural instinct either. You feel like you are missing out.

3. When you think about a new business unit, you see all the risks, all the potential things that could go wrong, and you don't launch or you do it in a half-hearted way. You tick all the right boxes, but you know you're doing it half-baked. To go 'all out' and 'all in' just seems too risky. Our instincts scream 'No!'

This is why there are so few exceptional leaders, so few stellar businesses, and so few Xcelerators. It's because we need to defy our natural instinct to turn away from the threat and instead do the opposite. As one of the few exceptional leaders, Jeff Bezos, puts it:

> What we need to do is always lean into the future; when the world changes around you and when it changes against you – what used to be a tailwind is now a headwind – you have to lean into that and figure out what to do because complaining isn't a strategy.

Companies don't have instincts. People do.

So the more I think about it, the more I realise the world is relying on *you*. We are relying on you to take the tools offered in this book, learn them and use them for the good of your team, your organisation and for you!

Even if things are imperfect, it's better to start than wait. One of my mentors taught me that once you gain awareness, you have a responsibility to act. Once you know something is

wrong or right, you can't unknow it. You can only ignore it. You have the data: company lifecycles are shortening. Rapid change is threatening traditional performance, threatening jobs, even the existence of many incumbents. It's coming hard after your comfort zone.

So it's up to *you* now. It's up to you to take over from here. It's up to you to start applying the Xcelerate framework, so that you can create the next big wave. Do something different now, before it's too late. There is help out there; you don't have to go it alone. It's just about starting.

Some final words of wisdom come from Reed Hastings, founder and CEO of Netflix, who said: 'Most entrepreneurial ideas will sound crazy, stupid and uneconomic, and then they'll turn out to be right.'

Now you're ready to go off to make some big moves of your own. But as you do, be sure to keep in touch and let me know how you get on.

I'm looking forward to having you in my tribe of seriously strategic leaders.

Paul
paul@paulbroadfoot.com.au

SOUR CES

PART X - UNDER ATTACK!

1. Walls don't work

Bose, P 2004. *Alexander the Great's Art of Strategy: Business Lessons from the Great Empire Builder*

Alexander's Siege of Tyre, 332 BCE, Ancient History Encyclopaedia, www.ancient.eu

Battering Rams, Medieval Warfare, www.medievalwarfare.info

Kurke, L 2004. *The Wisdom of Alexander the Great: Enduring Leadership Lessons From the Man Who Created an Empire*

Rock, D & Grant, H 2016. *Why Diverse Teams Are Smarter*, Harvard Business Review, www.hbr.org

Hunt, V, Layton, D & Prince, S 2015. *Why diversity matters*. McKinsey & Company, www.mckinsey.com

Hull, L 2004. MHQ. *Medieval Warfare: How to Capture a Castle with Siegecraft*

Smartphone OS Market Share 2016 Q2, IDC, www.idc.com

Price, R 2014, review of Keating, G 2013. *Netflixed: The Epic Battle for America's Eyeballs*

Battersby, L 2016. 'Foxtel rejigs products to compete against streaming services Stan, Netflix, Presto', *Sydney Morning Herald*, 6 September

Evans, P & Forth, P 2015. *Borge's Map: Navigating a world of digital disruption,* BCG Perspectives, www.bcgperspectives.com

Blank, S 2016. *Why Visionary CEOs Never Have Visionary Successors*, Harvard Business Review, www.hbr.org

Mochizuki, T 2016. 'Taiwan's Foxconn Completes Acquisition of Sharp', *Wall Street Journal*, August 13

Sharp history, www.sharp-world.com

Reeves, M, Levin, S, Ueda, D 2016. *The Biology of Corporate Survival,* BCG Perspectives www.bcgperspectives.com

Estrin, J 2015. 'In 1975, this Kodak employee invented the digital camera. His bosses made him hide it', *The Australian Financial Review*, August 13

Anthony, SD, Viguerie, PS & Waldeck, A 2016. *Corporate Longevity: Turbulence Ahead for Large Organizations*, Innosight, www.innosight.com

2. Australian research leaves a bitter taste

Department of Industry, Innovation & Science, www.industry.gov.au/Office-of-the-Chief-Economist

Kennedy, S 2016. *Grim reality of Aussie innovation*, www.innovationaus.com

3. King tides of change

Singer Sewing Machines 1865–1970, www.singersewinginfo.co.uk

Industrial Revolutions Inventions Timeline 1712–1942, www.storiesofusa.com

2016 U.S. Music Mid-Year Report, www.nielsen.com

Everett, S 2015. *The Rise of Smart Phones in Cambodia Challenges Social Norms*, The Asia Foundation, www.asiafoundation.org

Phong, K & Sola, J 2015. 'Mobile Phones and Internet in Cambodia 2015'. (Research study)

Kellner, T 2013. *Jet Engine Bracket from Indonesia Wins 3D Printing Challenge*. GE Reports, www.gereports.com

Carson, B 2016. *Apple's $1 billion investment in Didi came together in less than a month*, Business Insider Australia, www.businessinsider.com.au

Here are the top 10 most successful American companies, Fortune, www.fortune.com

Clemons, EK & Row, M 1988. *McKesson Drug Company: A Case Study of Economost – A Strategic Information System*

Kopytoff, V 2015. *McKesson: The healthcare tech giant you've probably never heard of*, Fortune, www.fortune.com

PART Y - STRATEGY

4. Business model

Malone, TW, Weill, P, Lai, RK, D'Urso, VT, Herman, G, Apel, TG, & Woerner, SL 2006. *Do Some Business Models Perform Better than Others?* www.seeit.mit.edu

About eHarmony, www.eharmony.com.au

Stinson, L 2014. 'How GE plans to act like a start-up and crowdsource breakthrough ideas', *Wired*, April 2014

GE jet engine bracket challenge. GrabCAD Community, www.grabcad.com

Redrup, Y 2016. 'Business in Heels creates new marketplace, XChange, for professional women', *Australian Financial Review*, August 28

Swartz, E 2015. *The transformation of GE: From 'We bring good things to life' to industrial machines in the cloud.* www.metratech.com

Melbourne Angels: www.melbourneangels.net

Kickstarter: www.kickstarter.com

TradeYa: www.tradeya.com

Xchange (Business in Heels): www.businessinheels.com

GrabCAD: www.grabcad.com

5. Be first and be lonely

Weber, J 1993. 'IBM Reports Largest Annual Corporate Loss: Earnings: The grim figures, though expected, put more pressure on company Chairman John Akers', *Los Angeles Times*, January 20

Gerstner Jr, LV 2002. *Who Says Elephants Can't Dance?*

Simoudis, E & Power, B 2016. *The 5 things IBM needs to do to win at AI,* Harvard Business Review. www.hbr.org

Barnett, M 2011. *IBM's Ginni Rometty: Growth and comfort do not coexist.* Fortune, www.fortune.com

Madrigal, AC 2011. 'IBM's First 100 Years: A Heavily Illustrated Timeline', *The Atlantic*, June 16

Chafkin, M 2016. 'IBM's First Female CEO on Why Bob Dylan is Talking to a Computer', *Bloomberg Business Week*, August 4

IBM Annual Report 2014, www.ibm.com

Langley, M 2015. 'Behind Ginni Rometty's Plan to Reboot IBM', *Wall Street Journal*, April 20

Brooks, J 2014. *Business Adventures: Twelve Classic Tales from the World of Wall Street*

Austin, RD & Nolan, RL 2000. *IBM Corporation Turnaround*, Case study 600-098. Harvard Business School

6. Revenue model

Business Chicks: www.businesschicks.com.au

Business Network International: www.bni.com.au

Van Damme, J & Stolk-Oele, M 2015. *8 Things You Probably Don't Know About Jet Engines,* www.news.com

Vitasek, K 2012. *The Rolls-Royce of Effective Performance-Based Collaboration*, Maintenance Technology, www.maintenancetechnology.com

Glader, P 2009. 'GE's Focus on Services Faces Test', *Wall Street Journal,* March 3

GE Aviation 2016. *GE Aviation Services Launches TrueChoice Product Suite,* www.geaviation.com

Chesbrough, H 2011. *Open Services Innovation,* Innovation Excellence, www.innovationexcellence.com

Karp, S 2007. *Google AdWords: A Brief History of Online Advertising Innovation*. www.publishing2.com

Price discrimination, Ecomomics Online, www.economicsonline.co.uk

Our history in depth. Google, www.google.com

7. All you can fly

Dunn, H 1976. *Back Home: A History of Citrus County, Florida*

Hartzell, ST 2002. *Voices of America – St. Petersburg: An Oral History*

Sharp, T 2012. *World's First Commercial Airline – The Greatest Moments in Flight*, SPACE.com, www.space.com

Surf Air: www.surfair.com

Airly: www.flyairly.com

OneGo: www.onego.com

Groetzinger, K 2016. *There's a new service offering unlimited flights on airlines like American, Delta, and JetBlue*, Quartz, www.qz.com

Freed, J 2016. 'Start-up Airly aims to disrupt how you fly from Sydney to Melbourne', *Sydney Morning Herald,* January 11

Chang, A 2016. 'Surf Air, the members-only airline, is poised for a growth spurt', *Los Angeles Times*, April 10

8. Communication model

Powell, R 2015. 'How Atlassian ignored 'smart people' to get to the top', *Sydney Morning Herald*, September 28

Tay, L 2015. *ATLASSIAN – THE UNTOLD STORY: How two Australian young guns built a $3 billion company, headed for an American IPO.* Business Insider Australia, www.businessinsider.com.au

Bass, D 2016. *This $5 Billion Dollar Software Company Has No Sales Staff,* Bloomberg Businessweek, www.bloomberg.com/businessweek

Kehoe, J 2015. 'Atlassian Prices IPO, Sets Record $US4.4b valuation', *Australian Financial Review*, December 10

Raichshtain, G 2014. *BSB Sales Benchmark Research Finds Some Pipeline Surprises,* Salesforce, www.salesforce.com

2016 DSN Global 100 List, Direct Selling News, www.directsellingnews.com

Flynn, K 2016. *Twitter update shows how fast companies respond to your complaints.* Mashable Australia, www.mashable.com

Kulp, P 2015. *US Companies ignore 80% of Twitter questions from*

customers, but they answer on Facebook, Mashable Australia, www.mashable.com

Cook, P, Henderson, M & Church, M 2012. *Conviction: How Thought Leaders Influence Commercial Conversations*

Global Trust in Advertising 2015, www.neilson.com

What Are the Biggest Challenges and Opportunities Facing PR Agencies now? Bulldog Reporter, www.bulldogreporter.com

Krausova, M 2015. *5 Customer Buying Trends You Can't Ignore*, CEB, www.cebglobal.com

The newest data on social customer care by Socialbakers, Socialbakers, www.socialbakers.com

Tortora, A 2016. *Meet Direct Selling's Billion Dollar Markets,* Direct Selling News, www.directsellingnews.com

Ozmorali, H 2016. *Global Direct Selling Industry in 2015,* The World of Direct Selling, www.theworldofdirectselling

9. Tinker, tailor, solder, spy

Pederson, JP 2004. *International Directory of Company Histories*, Vol.63

Dell History, Funding Universe, www.fundinguniverse.com

Michael Dell: Taking the direct approach. Entrepreneur, www.entrepreneur.com

Gale, T 2006. *Dell Computer Corp*, Encyclopedia, www.encyclopedia.com

'Dell's net revenue from 1996 to 2013', *Business Review Weekly,* Vol. 26, No. 12, 2013

Dell, M & Fredman, C 2006. *Direct from Dell: Strategies that Revolutionized an Industry*

10. Differentiation model

Peterson, H 2015. *McDonald's Australia reveals how the US are doing it all wrong.* Business Insider Australia, www.businessinsider.com.au

McDonald's opens first McCafe in U.S., Entrepreneur, www.entrepeneur.com

Ovans, A 2015. *What Is a Business Model?* Harvard Business Review, www.hbr.org

Porter, ME 1985. *Competitive Advantage: Creating and Sustaining Superior Performance*

Treacy, M & Wiersema, F 1993. *Customer Intimacy and Other Value Disciplines,* Harvard Business Review, www.hbr.org

Bradley, K & Hatherall, R 2013. *The powerful economics of customer loyalty in Australia,* Bain & Company, www.bain.com

Hagel. J & Singer, M 1999. *Unbundling the Corporation,* Harvard Business Review, www.hbr.org

Hurn, C 2012. *Stuffed Giraffe Shows What Customer Service Is All About,* Huffington Post, www.huffingtonpost.com

Seven Advantages of Employee Empowerment, The Ritz-Carlton Leadership Centre, www.ritzcarltonleadershipcenter.com

Google X: www.x.company

Muoio, D 2016. *Alphabet's 20 most ambitious moonshot projects,* Business Insider Australia, www.businessinsider.com.au

Bogle, A 2016. *If all goes to plan, Amazon might be the messiest tech story of 2017,* Mashable, www.mashable.com

Amazon Go: www.amazon.com

11. Wired for sound

Innovations/achievements, www.bose.com

Clynes, T 2013. *The Curious Genius of Amar Bose*, Australian Popular Science, www.popsci.com.au

Rifkin, G 2013. 'Amar G. Bose, acoustic engineer and inventor, dies at 83', *New York Times*, 12 July

Ferris, R 2016. CNBC. *How Amar Bose used research to build better speakers*, CNBC Make It, www.cnbc.com/makeit

The Speakers That Started It All, Bose Newsletter, *www.*bose.com

PART Z –ACTION

12. Diving in

Blank, S 2003, *The Four Steps to The Epiphany*

GP2U Telehealth Pty Ltd: www.gp2u.com.au

Griffith, C 2016. 'GP2U: Smartphone app brings the doctor to your phone', *The Australian,* March 24

Addady, M 2016. *Meet Ross, the World's First Robot Lawyer,* Fortune, www.fortune.com

Butcher, M 2016. *Goodbye accountants! Startup builds AI to automate all your accounting.* Techcrunch, www.techcrunch.com

Brown, T 2009. *Change by Design: How Design Thinking Transforms Organizations and Inspires Innovation*

Osterwalder, A & Pigneur, Y 2010. *Business Model Generation: A Handbook for Visionaries, Game Changers, and Challengers*

Ries, E 2011. *The Lean Startup: How Today's Entrepreneurs Use Continuous Innovation to Create Radically Successful Businesses*

13: Be a speeadboat, not a battleship

Brook, B 2016. *Dick Smith only the latest high street brand to bite the dust.* www.news.com.au

Warby Motorsport. *The world's fastest team on water,* www.warbymotorsport.com

Susan 2013. *The world's fastest boat & the race for the water speed record,* Boat Covers Direct, www.boatcoversdirect.com

McMahon, J 2013. *Warby aims to break his own record,* ABC Newcastle, www.abc.net.au

Ancestors of Kenneth. P. Warby, www.thetreeofus.net

Dunn, C 2013. *The Founders of A Nation: The First Fleet 1788,* Australian History Research, www.australianhistoryresearch.info

Hastings, R 2015. *How I Did It: Reed Hastings, Netflix,* Inc., www.inc.com/magazine

Carucci, R 2016. *Organizations Can't Change if Leaders Can't Change with Them,* Harvard Business Review, www.hbr.org

Aon Hewitt 2016. *2016 Trends in Global Employee Engagement*

Netflix Culture: Freedom & Responsibility. www.slideshare.net/reed2001

Shontell, A 2013. *Sheryl Sandberg: 'The Most Important Document Ever To Come Out Of The Valley',* Business Insider Australia, www.businessinsider.com.au

Laloux, F 2014. *Reinventing Organizations*

Hamel, G 2011. *First, Let's Fire All The Managers,* Harvard Business Review, www.hbr.org

Mazel, J 2014. *Flat and Fluid: How Companies Without Hierarchy Manage Themselves,* Medium, www.medium.com

Geraci, J 2016. *What I Learned from Trying to Innovate at the New York Times,* Harvard Business Review, www.hbr.org

Notes

Notes

Notes

'You don't wake up a game changer, you become one. Experience, expertise and passion are critical to fuel the pace of business innovation. Rigorous methods and a disciplined approach will maintain the necessary stamina in such a difficult journey. *Xcelerate* is the recipe to succeed in innovative transformation. It's a must-read for those who wish to play their part in disrupting their industry.'

Matthias de Ferrieres
CEO, Insurance Republic, Stark Group

'As we evolve in the 21st century we have no choice but to innovate beyond conventional thinking. Paul educates the market, but also pioneers a new way of thinking that will propel your company forward. I know from personal experience that he is not your average business consultant. Quite frankly there is **nothing** average about Paul. He writes with wisdom far beyond his years. So if you're interested in being leading edge then read this book and adopt the teachings. It will give you market advantage far beyond the norm.'

Sally Anderson
Founder/Director, Evolved Leadership

XCELERATE

'In today's climate of fear and uncertainty, never has it been more important to step up to the leadership plate, challenge old thinking and forge new ground. This book will give you the clarity, confidence and courage to do just that.'

Margie Warrell
Bestselling Author of Stop Playing Safe

'Bringing entrepreneurial approaches, which are more commonly used with startups, into the corporate innovation space can challenge the corporate mindset. *Xcelerate* offers leaders, who want to increase pace in this area, refreshing and new approaches to business model innovation and how to achieve successful launches for new growth units.'

Mike Herd
Executive Director, Sussex Innovation Centre, UK

'Growth is non linear, so we will need jumps in thinking around our business model, our sales model and our whole of business approach. This book sets a new high water mark as an approach to designing new business growth in rapidly changing times.'

Matt Church
Founder, Thought Leaders Global